The Successful Internship

TRANSFORMATION AND EMPOWERMENT IN EXPERIENTIAL LEARNING

SECOND EDITION

H. Frederick Sweitzer
University of Hartford

Mary A. King
Fitchburg State College

THOMSON
★
BROOKS/COLE

Australia • Canada • Mexico • Singapore • Spain
United Kingdom • United States

THOMSON

BROOKS/COLE

Publisher/Executive Editor: Lisa Gebo
Assistant Editor: Alma Dea Michelena
Editorial Assistant: Sheila Walsh
Technology Project Manager: Barry Connolly
Marketing Manager: Caroline Concilla
Marketing Assistant: Mary Ho
Project Manager, Editorial Production:
    Rita Jaramillo
Print/Media Buyer: Doreen Suruki

Permissions Editor: Joohee Lee
Production Service: Publishing Support
    Services
Copy Editor: Pam Suwinsky
Cover Designer: Andy Norris
Cover Artist: Britt Howe
Cover Printer: Webcom
Compositor: TBH Typecast
Printer: Webcom

For more information about our products,
contact us at:
Thomson Learning Academic Resource Center
1-800-423-0563
For permission to use material from this text,
contact us by:
Phone: 1-800-730-2214
Fax: 1-800-730-2215
Web: http://www.thomsonrights.com

Library of Congress Control Number: 2003105088
ISBN 0-534-55879-8

Brooks/Cole—Thomson Learning
10 Davis Drive
Belmont, CA 94002
USA

Asia
Thomson Learning
5 Shenton Way #01-01
UIC Building
Singapore 068808

Australia/New Zealand
Thomson Learning
102 Dodds Street
Southbank, Victoria 3006
Australia

Canada
Nelson
1120 Birchmount Road
Toronto, Ontario M1K 5G4
Canada

Europe/Middle East/Africa
Thomson Learning
High Holborn House
50/51 Bedford Row
London WC1R 4LR
United Kingdom

Latin America
Thomson Learning
Seneca, 53
Colonia Polanco
11560 Mexico D.F.
Mexico

Spain/Portugal
Paraninfo
Calle/Magallanes, 25
28015 Madrid, Spain

This book is dedicated, with love

From Fred
*To my father, Skip Sweitzer, who taught me to be a man,
to love simply and completely, and to give back more
than you take, lessons I am still trying to learn.*

From Mary
*To my father, Charlie King, who taught me
the importance of living, loving, and teaching from the heart,
and to my son, Patrick Zimmermann, a natural teacher,
my best ever, full of heart and soul.*

The Koru

The Koru is a symbol used by the Maori culture
in New Zealand to represent new beginnings, growth,
and harmony. This symbol also represents a fern
slowly unfolding toward the light.

Contents

Foreword

There's something in this book for every intern, from those who don't know what to expect or how to play a role in shaping their experiences to seasoned interns who want to get the most out of their internships. The key for any intern is to understand how to identify and especially how to create a quality internship experience. Interns should be able to recognize the potential even in the most challenging and difficult situations they may face. Having learning goals and objectives, a self-study plan, reflective self-evaluations, and goals are essential to making the most of the experience.

The authors of this book have laid out some of the most critical issues of a quality internship experience. While directed primarily toward the student, it is also an essential guide for the internship professional. I would recommend they read this book together, so they both understand how to make this form of experiential learning a valid, productive experience. An internship needs to achieve a balance between learning and contributing in a supervised setting. Interpersonal relationships, reflection, self-understanding, and feedback are essential factors in creating that balance. This book places these in context and helps both parties achieve the maximum benefit from the internship.

Read this book carefully. You will see how to work within the system and within the culture of internships. You'll learn how to evaluate your situation in a realistic way. Follow the book's advice, and who knows, you may be hired and wind up supervising the next generation of interns.

Eugene J. Alpert, Ph.D.
President, National Society for Experiential Education
Senior Vice President, Washington Center
    for Internships and Academic Seminars
February 2003

Preface

When we first began to write about internships in 1994, our primary audience was human service education programs. There were very few books or articles at the time that addressed the experiences and needs of faculty and students in those programs, in contrast to the resources available to other helping professions, such as counseling and social work. We wanted to fill that gap. Since then the literature on human service internships has grown. We have found additional audiences for our work, including other helping professions, academic programs outside the behavioral sciences, and the emerging area of academic service-learning. The core of our thinking has remained the same over time, yet has grown and changed in response to the new audiences and ongoing research in the field.

We have intended this book to be more phenomenological than skill-based. It is not about the terminology, skills, and knowledge that students need in order to work with a particular population or in a particular setting. Rather, it is about the *experience* of the internship, the evolving process of self- knowledge, and the confrontation and overcoming of barriers to growth. The ideas, theories, and exercises are designed to help interns make sense of their experience and their reactions to it every day, every week. These, we have found, are useful and vital across varied contexts and settings.

The book is organized around two "big ideas." The first is that there are issues and concerns that students encounter at certain stages in their experiences in the field. Students progress through these developmental stages in a reasonably predictable order, though not at any predictable rate of speed. The second big idea is the importance of self-understanding. In addition to knowing about clients, techniques, methods, organizations, communities, and administration, students need to know something about themselves. We believe that an enhanced understanding of themselves makes them better practitioners, citizens, and people. It also helps them recognize and make use of the developmental stages, because individuals experience the stages in their own unique ways.

We expect much diversity in the field experiences of the students and faculty who make use of this book. In some cases, the students will be in a culminating internship, and most of the specific skill development will have been accomplished earlier in their programs. In other cases, field experiences are interwoven throughout the entire program. Some students will use this book as part of an on-campus or online seminar (that is how we use it); others will be using it as a self-guide through the field experience, communicating with instructors individually but not meeting with peers or faculty in groups.

Because of this diversity, we have tried to give students and instructors as much flexibility as possible in using the book. We have included resources for further exploration so that students and instructors can build on those areas that are of interest and relevance. We have also provided a wide range of reflective questions at the end of each chapter, anticipating that students and instructors will choose from among them. Some of those questions work best for individual reflection and some are designed for group reflection and discussion. We have also tried to create a book that can be augmented or supplemented with more skill-specific books, articles, and instruction.

Our experience has been that students go through the stages in the order we discuss, but often at different rates of speed. In our seminar classes, we cover the chapters in order and at a particular time in the internship, and that seems to work for the majority of the group. However, we also listen carefully to what individual students are saying in class and in their journals, and we often suggest that a student reread a chapter or skip ahead to future ones as needed.

The chapters do not need to be read in the order in which they are written, but some caveats are in order. Chapters 1–4 are sometimes read before the internship or field experience begins, or as part of a prerequisite seminar. Some instructors will want to have students read the chapter on legal and ethical issues much earlier. Others may want to have students read and think about colleagues before clients. Still others, depending on the context, may want to have students consider the community or organizational context before turning to consideration of either clients or colleagues. We do suggest, though, that the order of the stages be preserved. Chapters 5–9 pertain to the Anticipation stage, and we recommend reading them, in whatever order seems appropriate, before moving on. Chapters 10 and 11 should come next, and speak to the concerns of the next two stages: Disillusionment and Confrontation. Chapters 12 and 13 pertain to the Competence stage. And Chapter 14 is designed to help students think about the ending of the internship. It should be read, though, in enough time for students to anticipate and plan for the concerns and issues of the Culmination stage.

NEW FOR THIS EDITION

Throughout this edition we have tried to make the voices of students more prominent. Quotes from journals, papers, exercises, and e-mails appear frequently. Each chapter of the book has been updated and augmented, but some specific changes are worth noting:

- Chapter 1 features increased coverage on reflective techniques and conducting an effective seminar.
- Chapter 5 features expanded treatment of the learning contract.
- In the first edition, a chapter called "Getting to Know the People" addressed clients, supervisors, and coworkers. In this edition, clients have their own chapter, with a new section on personal safety.
- Chapter 8, on the organizational context, has been reorganized around Bolman's and Deal's organizational frames.
- We have included a new chapter (Chapter 9) on the community. Some students are involved in experiences where the primary focus of the work and the intended

change is not an individual, or even a group, but a community. In addition, we believe that some knowledge of community dynamics is important for any setting.

- Ethical and legal issues, once a section in a chapter, have been given their own chapter (Chapter 13). The treatment of these issues has been greatly expanded.

ACKNOWLEDGMENTS

This edition of *The Successful Internship* captures the growth and development of our thinking of the past five years. And that growth and thinking has been done in the context and company of wonderful students, valued colleagues and mentors, and an invaluable network of family and friends. At the risk of forgetting someone, let us try to acknowledge this vibrant support system.

Our colleagues have listened to our ideas, used our book, and expanded our thinking with their ideas. We thank in particular Heather Lagace at the University of Hartford, faculty member in human services, whose energy, creativity, and enthusiasm for our book and for working with interns parallels our own.

The National Organization for Human Service Education, the New England Organization of Human Services Education, and the National Society for Experiential Education have been our professional homes and have provided us with forums to present and discuss our ideas. We thank our many colleagues in those organizations who have attended workshops, shared ideas, and reviewed and published our work. Our special thanks go to Peggy Anderson, Mark Homan, Will Holton, Lynne Kellner, Pat Kenyon, Tricia McClam, Lynn McKinney, Trula Nichols, Diane McMillen, Mary Davidson, and Marianne Woodside, Lynn Gaulin, Al Cabral, Dierdre Simpson, Gene Alpert, Roberta Magarrell, and Dwight Giles, who have shown particular interest in and support for our work.

Many reviewers contributed their time and expertise to making this a better book. Deborah Altus, Washburn University; Peggy Anderson, Western Washington University; Montserrat Casado, University of Central Florida; Marcia Harrigan, Virginia Commonwealth University; Christopher Rybak, Bradley University; Tom Struzick, University of Alabama; and Stephen Wester, University of Wisconsin, Milwaukee, provided thorough reviews, astute insights, and valuable suggestions at various stages in the writing process. In addition, Mark Homan of Pima Community College was kind enough to help with the chapter on communities, and Steve Eisenstat of Suffolk University Law School was generous in his assistance with the chapter on legal issues. Our heartfelt thanks go to each of them.

Our students continue to be our most powerful teachers. It is their experiences in the field and the classroom that have shaped this edition most powerfully. Their dedication, courage, resilience, and diversity keep us coming back, year after year.

We are grateful as well for the support of agency supervisors who have cared about and for our students, for their insight and their unswerving commitment to quality internships. Particular thanks go to Tim Blair, Chris Garahan, Ken Gwozdz, April Morrison, Joe Pedemonte, Rose Santiago, Ed Denmark, Henry Culver, and Wes Colter.

Without the technical assistance of a number of people, we would not have been able to bring our work to fruition. We are most appreciative for the help we were given by Karen Rogers, Catherine Ricketts, Walter Beaman, Joan MacDonald, and Linda Medeiros; and for the ongoing sleuth work of Janice Ouellette, librarian extraordinaire at Fitchburg State College.

Our appreciation goes to the staff at Brooks/Cole. Alma Dea Michelena kept us on track and on schedule throughout the process. In addition, we thank Rita Jaramillo, project manager; Vicki Moran, production service; Pam Suwinsky, copyeditor; and Joohee Lee, permissions editor. Special thanks to Vern Boes and Britt Howe for the cover.

As always, our families and friends have showered us with support, interest, and attention during this project. Family members Patrick Zimmermann, Skip and Betty Sweitzer, Sally Sweitzer and Britt Howe, Charlie King, Phyllis Agurkis, Joe King, and Heidi Sadowski deserve special mention. So, too, do our dear friends Jeff and Judy Bauman, Vicky Day, Regina Miller, Kathy Rondeau, Mary Ann Hanley, Sam Pukitis, Steve Eisenstat, Vivien Perge, Ed Weinswig, and the Magnificent Six, a wellspring of inspiration.

Finally, our partners in this endeavor and all of life's undertakings, Martha Sandefer and Peter Zimmermann, get an extra measure of thanks and love for their patience, assistance, and understanding throughout this edition, and, now that it is finished, an extra measure of our attention!

ABOUT THE AUTHORS

H. Frederick Sweitzer is Professor of Human Services at the University of Hartford in Connecticut, where he also serves as Associate Dean of the College of Education, Nursing and Health Professions. Fred has more than 25 years' experience in human services as a social worker, administrator, teacher, and consultant. He has placed and supervised undergraduate interns for 18 years and developed the internship seminar at the University of Hartford. Fred brings to his work a strong background in self-understanding, human development, experiential education, and group dynamics. He is on the editorial board for the journal *Human Service Education* and has published widely in the field. Fred has held a variety of positions on the board of the National Organization for Human Service Education.

Mary A. King is Professor, Behavioral Sciences Department, at Fitchburg State College, where she instructs seminar and colloquium classes for interns and service-learning students. She has held administrative positions in academic programs and has placed and supervised interns from a variety of academic majors. Mary has more than 30 years' experience in the helping professions and in experiential education. She has published in these fields and holds several professional licenses. Mary brings to her academic work a background in public education, juvenile justice, clinical counseling, and consultation. She has served on regional and national boards in human service education, local boards of human service agencies, and presently serves on the board of the National Society of Experiential Education.

BEGINNING
THE JOURNEY

CHAPTER 1

Laying the Groundwork

Education is revelation that affects the individual.
GOTTHOLD EPHRAIM LESSING, 1780

It gives meaning to everything you have learned and makes practical sense of something you've known as theoretical.
STUDENT REFLECTION

WELCOME TO YOUR INTERNSHIP

You are beginning what is, for most students, the most exciting experience of one's education program. Chances are you have looked forward to this experience for a long time. You've probably heard your share of stories—both good and bad—from other, more experienced students. And while, as an intern, you may be part of a small minority on your campus, you join virtually thousands of other students all over the country. An intensive field experience is a critical component of many education programs—including psychology, sociology, and criminal justice—and of almost every human service education program (E. Simon, 1989). Many human services students conduct their internships in a social service setting; others choose corporate or research settings.

In addition to internships required by certain majors, there are increasing numbers of colleges and universities requiring or encouraging an intensive community service experience regardless of major. Van Grinsen (1995) points out that in 1979 one student in 36 participated in an internship, while in 1995 the number had climbed to one in five. The number of hours required varies widely, from as few as 80 hours to as many as 500 (E. Simon, 1989). There are many terms for these experiences, including *internship, field work, field experience, co-op education, field education*, and *practicum*, and no consensus on the meaning of any of those terms. We use the term *internship* to refer to those experiences that involve receiving academic credit for working in a social service agency, or at some other site, for at least 8 hours per week over the course of a semester.

Also, there are increasing numbers of professors and programs that are integrating service-learning into their courses. Service-learning involves working in the community, but it goes beyond volunteering and emphasizes specific academic gains related to the objectives of the course. While these are not internships as we have defined the term, much of what is in this book will be useful to students and professors in service-learning settings.

The internship is the culminating academic experience for many students; some programs have several internships, but for many students the internship occurs near the end of the academic program and can be a chance to pull together and apply much of what you have learned. Many students see the internship as a chance (finally) to learn to actually *do* something. Although you may have had courses that developed particular skills, the internship is a chance to improve those skills and acquire lots of new ones.

Skill development, though, is only one of the possible goals and outcomes of your experience. An internship offers a chance to apply theory to practice. Actually, it is a chance to develop the relationship between theory and practice, for each should inform the other (Sgroi & Ryniker, 2002). The theories you have studied (and continue to study) should help you analyze and perform effectively in various situations. However, your experience will also help you see where the theories do not quite apply or where you need to search for a new theoretical model to help you. Thus, theories are transformed through their application.

The internship also affords you the opportunity to understand the world of work in a more complete way than you do now. Even if you have had full-time jobs, presumably your internship is taking you into an area in which you have little professional experience.

The internship is also a catalyst for personal growth. If you give yourself a chance, you can learn a tremendous amount about yourself. Finally, the internship can help you clarify your career and educational goals.

So Why Do You Need a Book?

The internship is a learning experience like no other. In a classroom learning experience, you learn through readings, lectures, discussions, and exercises. These are the raw materials you are given. You bring your ability to memorize, analyze, synthesize, and evaluate to these materials—these are your learning tools. The arena for learning is the classroom, and classrooms vary in the amount and quality of interaction between students and the instructor, and among students. Some can be very interactive places, with mutual dialogue among students as well as between students and teachers. Other classrooms are interactive, but only between teacher and student; it is almost as if there is a multitude of individual relationships being carried on in isolation. The internship is different in every way, and this book is designed to help you and your instructor with this new approach.

Furthermore, the internship is not just an intellectual experience. It is a *human* experience, full of all the wonderful and terrible feelings that people bring to their interactions and struggles. And your clients are not the only ones having these feelings; you will have them too. This emotional, human side of the internship is not just a backdrop to the real work and the real learning; it is every bit as real and important.

Experience, both intellectual and emotional, is the raw material of the internship. You will be learning mostly through experience, although you may engage in some traditional academic activities as well. However, one noted theorist in the field of experiential education, David Kolb (1984), suggests that experience alone does not lead to learning or growth. Rather, the experience must be processed and organized in some way. You must think about your experience, sometimes in structured ways, and discuss it with others. Reflective dialogue with yourself and your peers is the primary tool for learning in the internship. This book is designed to help you structure that reflection and dialogue. It will invite you to think about your internship in a variety of ways, some of which may be new to you. It may also help you anticipate some of the challenges that await you and move successfully through them.

Relationships are the medium of the internship; they are the context in which most of your learning and growth occur. Many of you will have clients at your placement, but even if you do not, you will be involved in relationships with a supervisor at your placement, an instructor on campus, other interns on-site and on campus, and coworkers at your placement. These relationships offer rich and varied opportunities for learning and growth. This book will help you think about these relationships, capitalize on the opportunities they present, and address any problems that may arise.

In addition to excitement and satisfaction, most interns also experience some real difficulties—moments when they question themselves, their career choices, their placements, or all of these. We like to think of these moments as crises, but not in the way you are probably familiar with the term. We prefer to think of crises as the Chinese do. The symbol for *crisis* in Chinese is a combination of two symbols: danger and opportunity (Figure 1.1).

So, while there is some risk and certainly some discomfort in these challenges, there is also tremendous opportunity for growth. We hope this book will help you see both the dangers and the opportunities inherent in an internship and grow from both. We will encourage you not to run from these crises, but to meet them head on, with your mind and your heart open to the experience.

The truth is that some of you may not need a book; you will learn all you can and survive the crises just fine. In other cases, you will want to use this book as a self-guiding tool, without a formal class, and we have tried to write it so that you can do that.

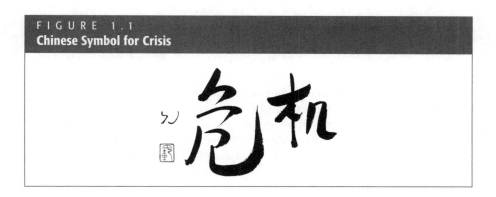

FIGURE 1.1
Chinese Symbol for Crisis

However, we believe that for most students, this book, in combination with a skilled instructor, supportive peers, and some of your time and energy, can add to your learning.

We first conceived of this book for a human service internship audience, but there is an incredible variety of programs and placements that fall under that umbrella. Add to that the audiences in other professions such as counseling, social work, nursing and business, and service-learning, community service, and other field-based experiences, and the variety becomes even wider. So, too, is the amount of preparation that students bring to these experiences. This book is meant as a guide to the phenomenological experience of the internship, to help you anticipate and make sense of the emotional aspect of your work. But because of the variety we just mentioned, it is very important that you consult your guiding faculty member early and often, even if there is no formal class that accompanies your experience. Decisions about what theories to explore and what parts of this book to emphasize should be made by the intern and faculty member together. Decisions also need to be made about your skill sets and places where they may need some attention. We are speaking here of specific skills that are relevant to specific placements and fields, but also of more generic skills, such as communication and group process skills. We urge you to be proactive in identifying these areas for growth and planning to attend to them.

Some Basic Terms

Although internships, co-op education, community service, and service-learning experiences exist at many colleges and universities, different language is often used to describe the various aspects of the experience and the people associated with it. For example, the term *supervisor* sometimes refers to a person employed by the placement site and sometimes to a faculty or staff member at the college. So, at the risk of boring those of you who feel very clear about these terms, we take a moment now to be clear about what we mean.

- **Placement or Site**   This term refers to the place where the student is working, and sites can vary quite a bit. It could be a social service agency, a corporate setting, a college or university office, a hospital, or a school. Through the process of finding a placement, you probably are aware of the incredible variety of opportunities that exist in the community. If not, and if you are curious, there are books mentioned at the end of this chapter that you can consult.

- **Intern**   This is the term that refers to you, the student who is working at the site, even though you may not be called an intern at your college or university. If you are in a community service or service-learning activity, you will probably be referred to as a volunteer or a student.

- **Supervisor**   Your supervisor is the person assigned by the placement site to meet regularly with you, answer your questions, and give you feedback on your progress. Most placements assign one site supervisor to one student, although in some cases, there may be more than one person fulfilling these functions. Some academic programs use the term *field instructor* to describe this person in order to emphasize the educational (as opposed to managerial) nature of the role.

- **Instructor**   This is the faculty or staff member at your school who oversees your placement. In some cases, it is the same person who helped you find the placement, but in other cases, it is someone else. This person may meet with you individually during the semester, visit you at the site, hold conferences with you and your supervisor, conduct a seminar class for interns, grade your performance, or handle all of these tasks.

- **Coworker**   This term refers to the other people who work at the placement, regardless of their title, status, or how much you interact with them. If there are other students at the site, from your school or some other school, they are in the role of coworker when you are at the placement site.

THE CONCEPTS UNDERLYING THIS BOOK

Experiential Education

An internship, community service, or service-learning experience, like other kinds of field instruction, is a form of experiential education. While this approach to learning may not be well understood in many places on your campus, it comes out of a long theoretical and practical tradition, and we thought you might want to know something about it.

Experiential learning has philosophical roots dating back to the guild and apprenticeship systems of medieval times through the Industrial Revolution. Toward the end of the nineteenth century, professional schools required direct and practical experiences as integral components of the academic programs (such as medical schools and hospital internships, law schools and moot courts and clerkships, normal schools and practice teaching, forestry/agriculture and field work) (Chickering, 1977).

Perhaps the best-known proponent of experiential education was the educational philosopher John Dewey (1916/1944, 1933, 1938/1963, 1940). Dewey believed strongly that "an ounce of experience is better than a ton of theory simply because it is only in experience that any theory has vital and verifiable significance" (1916/1944, p. 144). However, he was convinced that even though all real education comes through experience, not all experience is necessarily educative. This idea was reiterated by David Kolb (1984, 1985; Kolb & Fry, 1975), whose work was mentioned earlier and who emphasized along with Dewey the need for experience to be organized and processed in some way to facilitate learning. Dewey also felt strongly that the educational environment needs to actively stimulate the student's development, and it does so through genuine and resolvable problems or conflicts that the student must confront with active thinking in order to grow and learn through the experience.

Experiential education is based on the premise that for real learning to happen, students need to be active participants in the learning process rather than passive recipients of information given by a teacher. When learning is a passive process, teachers are the centers of energy who tell you the information that they think you need to know. But when learning is an active process, students are the centers of energy. The teacher's role is to guide or facilitate your learning by taking an interest in your work and coaching you through the experience (Garvin, 1991). As an active participant in the learning process, you play a central role in shaping the content, direction, and pace of your learning.

Predictable Stages

This particular book, while grounded in experiential education theory, is based on two "big ideas." The first is that interns go through predictable stages of development during the course of the internship. Over the years, as we have supervised interns and worked with students engaged in service-learning, listened to their concerns and read their journals, talked with their site supervisors, and discussed similar experiences with colleagues and students at other institutions, a predictable progression of concerns and challenges began to emerge. We have organized these concerns into stages, which are modeled after the work of Lacoursiere and Schutz (Lacoursiere, 1980; Schutz, 1967; Sweitzer & King, 1994). Understanding this progression of concerns will help you, your instructor, and your site supervisor predict and make sense of some of the things that may happen during the course of your placement and think in advance about how to respond. It will also help you view many of your thoughts, feelings, and reactions as normal, and even necessary. The experience then becomes a bit less mysterious, and for some people that makes it more comfortable.

For example, if you are feeling excited but also somewhat anxious as you begin the placement, you may wonder whether that anxiety is a sign of trouble or where it may have come from. Knowing that it is a common and predictable experience will help you stop worrying about the fact that you are anxious and let you direct your energy toward moving through that anxiety. Here is what one of our students had to say:

> Now that I know I am not the only one that is concerned with these feelings
> I am better able to share them with others without feeling embarrassed.

Self-Understanding

The second major idea behind this book is that to make sense of your internship or service-learning experience, you need to understand more than a stage theory. You need to understand yourself. No two students have the same experience, even if they are working at the same agency. That is because any internship experience is the result of a complex interaction between the individuals and groups that make up the placement site and each individual intern. You are a unique individual, and that uniqueness influences both how people react to you and how you react to people and situations. You view the world through a set of lenses that are yours alone. Therefore, each of you will go through these stages at your own pace and in your own way. Events that trouble you may not trouble your peers, and vice versa. Some of you will be very visible and dramatic in both your trials and your tribulations. Others will experience changes more subtly and express them more quietly.

Finally, no one experiences an internship in a vacuum. You have a life outside the placement (although it may not seem like it sometimes), and your network of family, friends, and academic and professional obligations will shape your experience in a powerful way. We want to help you think about yourself throughout your internship in ways that we believe will lead you to important insights and to a smoother journey on your path to personal and professional development.

In summary, then, the stages of an internship will help you understand internships in general and some of the things that are liable to happen during your internship. Understanding yourself will help you recognize the particular style in which you will experience the internship. Putting both of those pieces of knowledge together will give you a powerful tool to understand what is happening to you, to meet and deal successfully with the challenges you face, and to take a proactive stance in making your internship the most rewarding experience it can be.

TOOLS FOR LEARNING: REFLECTION AND SEMINARS

To take advantage of these two big ideas and to maximize your learning during your internship, you need some basic tools. These tools may be somewhat different from the ones that have brought you success—or struggle—in other kinds of learning experiences. As an individual, you need to develop the skills and the habit of reflection. And as a member of a seminar group, you need to learn to use that class to promote everyone's learning and facilitate their journeys through the stages of an internship.

Reflection

Reflection is a fundamental concept in experiential education. In order to turn your experience into learning you need to stop, recall events, analyze and process them. Dwight Giles, who has written extensively about service-learning and internships, makes a point about service-learning that we think applies to all the experiences covered in this book (Eyler & Giles, 1999; Giles, 2002). He says that reflection is what connects and integrates the service, or the work in the field, to the learning. Otherwise, whatever theory you study can be emphasized in your classes but not necessarily integrated with practical experience. At the other extreme, practical experience is left to stand on its own. Reflection is the connection, and it is a powerful key to your success, your growth, your learning, and even your transformation.

While this may sound daunting, you actually do it all the time. If you are walking back from class and find yourself mulling over the remarks of a professor, or wondering how a classmate came up with a particularly interesting comment, you are reflecting. If you are in the car and start to think over an argument you have had with your child, your partner, or a friend, trying to figure out what happened and how you could have handled it differently, you are reflecting.

These examples, though, are instances of spontaneous reflection; we want you to make reflection a deliberate and regular habit. *Reflection* means to look back, and there will come a time when we ask you to look back on your experience as a whole, but for your internship the process should start at the beginning and be integral to the process (National Society for Experiential Education, 1998) Developing the habit of productive reflection takes patience, practice, and discipline. It means setting aside quiet time to think, because as one of our students put it, "The best answers come from the silence within." And it means resisting the temptation to just keep going from one activity to the other, in your internship and in your busy life; another intern said, "The internship proceeded at such a fast pace that I often felt it was one step ahead of me." Paradoxically, we have found that one of the best ways to stop the internship from getting ahead of you is to make time to stop and think.

There are lots of techniques for reflection. We will discuss some specific ones but it is not a comprehensive list; you will need to find the one or ones that work best for you, your instructor, and your situation. Remember, too, that there is a difference between reflecting and recording, although the two can overlap. Your instructor, your placement site, or both, may have specific ways that they want you to keep a record of what you have done. These records may be used for documentation and be kept in official files. They may be used in supervision as well. The primary purpose of reflection, though, is to promote your growth and learning, and the primary audience for a reflective technique is you. Eyler and Giles (1999) have offered the "Five Cs" of reflection that can guide you in selecting and assessing potential techniques. The Five Cs are Continuous, Connected, Challenging, Context, and Coaching. In keeping with these principles, you need to make reflection a habit; structure and connect that reflection to learning goals; and make sure you are challenged to reflect more deeply and through a wider range of lenses. You also need to work with your instructor to see that that your choice of techniques makes sense for your particular context, and that you get the coaching and guidance you need to use any technique to your best advantage.

KEEPING A JOURNAL

> *I believe that classwork and journals are critical to internships because they allow support from peers, feedback from teachers, and reflection on your own work and feelings.*
>   STUDENT REFLECTION

One of the most powerful tools for reflection that we know of is keeping a journal. Your instructor may require you to keep a journal of some kind, but even if it is not required in your setting, we strongly recommend that you keep one. We also suggest that you write an entry at the end of every day that you go to your internship. Although it may occasionally seem like a chore, if you put time into it, journal keeping will give you a way to see yourself growing and changing. It also forces you to take time on a regular basis to reflect on what you are doing. Many of the quotes you have seen and will continue to see throughout this book are drawn from student journals. A well-kept journal is a gold mine to be drawn on for years to come. It becomes a portfolio of the experience as well as a record of the journey.

Again, perhaps the most important thing you can do for your journal is to allot sufficient time to do it. Doing it over lunch on the due date is not a good approach! For many of you it is going to take practice and focus to learn to write in your journals in the most effective and productive way. As you plan your days and weeks, leave at least 30 minutes at the end of each day at your internship to write.

If you have a disability that makes it impossible or difficult for you to write, or if writing does not come easily to you, your journal could be tape-recorded instead. Your instructor can listen to the tape each week and respond to you on tape or in writing, whichever the two of you prefer. Of course, you will need to negotiate these arrangements with your instructor, but a little time and thought should yield a method that allows you to reflect comfortably on your experience and maintain a dialogue with your instructor.

For those of you who are doing your internship at a great distance from campus, or as part of a distance learning program, the journal is even more important. In addition to the benefits already mentioned, the journal and responses to it are a way for you and your campus instructor to have a continuing conversation about your work and your reactions to it. Advances in Web-based technology such as Blackboard and WebCT make it easy to send journal entries back and forth. If you do not have access to these technologies, e-mail can work just fine.

If you do decide to keep a journal, make sure you are very clear with your instructor, supervisor, and clients about the intent of the journal and issues of confidentiality. If your journal is for your personal use only, then there is no issue. You have full responsibility for its contents and for ensuring that what you write is for your eyes only. However, if you want, or are required, to show it to other interns, your instructor, your supervisor, or anyone else, you must be careful not to disclose information about clients, the placement, or even yourself that is supposed to be kept private. Discuss this issue with your instructor and your site supervisor before going too far with your journal. You may also be concerned that you cannot be completely candid in your journal if some of the people you are writing about are going to read it! Some interns keep their journals in loose-leaf format and merely remove any pages they wish to keep private before showing the journal to anyone else.

There are many different approaches to journal writing, and many different reflective techniques. Your instructor may have forms and techniques that you are required to use, but we would like to discuss just a few of the more common forms here.

Unstructured Journals   The simplest form of journal writing is just to take time after each day to think back on what stood out for you that day. Although there is no "right" length for these entries, they should record what you did and saw that day, new ideas and concepts you were exposed to and how you can use them, and your personal thoughts and feelings about what is happening to you. It may be helpful to divide what you learn at an internship into four categories: (a) knowledge, (b) skills, (c) personal growth, and (d) career development.

Knowledge refers to things you know about; for example, you might learn the principles of behavior modification. *Skills* are things you know how to do; for example, you know how to set up a behavior management program. *Personal growth* refers to what you have learned about yourself and your attitudes, values, reaction patterns, and personality traits. *Career development* refers to what you are learning about the field of human services, counseling, social work, criminal justice, or whatever profession you are considering, and your place in it. Try to include all these categories in your journal.

Many interns tell us they are afraid that there are going to be days when there is just nothing to say. Well, our experience is that you won't have that happen very often, but there may be some days when writing is difficult. For those days, here are some questions to consider, generated by a community service program (National Crime Prevention Council, 1988):

WHEN YOU DON'T KNOW WHAT TO WRITE

- What was the best thing that happened today at your site? How did it make you feel?
- What thing(s) did you like least today about your site?
- What compliments did you receive today and how did they make you feel?
- What criticisms, if any, did you receive and how did you react to them?
- How have you changed or grown since you began your work at this site? What have you learned about yourself and the people you work with?
- How does working at this site make you feel? Happy? Proud? Bored? Why do you feel this way?
- Has this experience made you think about possible careers in this field?
- What kind of new skills have you learned since beginning to work at this site? How might they help you?
- What are some of the advantages or disadvantages of working at this occupation?
- If you were in charge of the site, what changes would you make?
- How has your work changed since you first started? Have you been given more responsibility? Has your daily routine changed at all?
- What do you think is your main contribution to the site?
- How do the people you work with treat you? How does it make you feel?
- What have you done this week that makes you proud?
- Has this experience been a rewarding one for you? Why or why not?

> FROM "REACHING OUT: SCHOOL-BASED COMMUNITY SERVICE PROGRAMS," BY NATIONAL CRIME PREVENTION COUNCIL, 1988, P. 101. REPRINTED WITH PERMISSION.

Other Kinds of Journals   Some other forms of journals that have been used with interns include these (Baird, 2002; Bringle & Hatcher, 1999; Inkster & Ross, 1998):

- **Key Phrase** journals are those in which you are asked to identify certain key terms or phrases as you see them in your daily experience.
- **Double Entry** journals are divided into two columns. In one column you record what is happening and your reactions to it. In the other you record any ideas and concepts from classes or readings that pertain to what you have seen and experienced.
- In **Critical Incident** journals you identify one incident that stands out over the course of a day, or a week, and write about it in some depth.

PROCESSING TECHNIQUES

There are a number of processing techniques that you can use and that you may want to include in your journal. Some techniques are specific to a discipline, such as "verbatims" in pastoral counseling or "process recording" in social work. If you are not in one of these fields, you can use different ones at different times, or you can use one consistently depending on your preference or that of your instructor. There are merits to both choices. Switching techniques from time to time may let you see things you have been missing. On the other hand, using a consistent technique allows you to look back over several entries and look for patterns.

Three-Column Processing   Gerald Weinstein (1981; Weinstein, Hardin, & Weinstein, 1975) developed a method of reflecting on events that may be helpful with your journal. Take a moment at the end of the week to recall any events that stand out in your mind. Select one or two (they can be positive or negative). Divide a piece of paper into three columns. In the left-hand column, record each action taken by you or others during the event.

Record only those things that you saw or heard, such as "She frowned," "He said thank you," or "They stomped out of the room." List them one at a time. Now, review the list and try to recall what you were thinking when the different actions occurred. When you recall something, enter it in the middle column, directly across from the event. For example, you may have been thinking, "What did I do now?" when the people left the room. Finally, read the list again and try to recall what you were feeling at the time each action and thought occurred. Record what you recall in the right-hand column. For example, you may have felt embarrassed, confused, or angry when they walked out. Table 1.1 presents an example of this sort of analysis.

The Integrative Processing Model   Pam Kiser has developed another reflective technique for interns called the Integrative Processing Model (Kiser, 1998, 2000), consisting of six steps:

1. **Gathering Objective Data from Concrete Experience**   In this step you select an experience that you have seen or been part of. You can use a written, videotaped, or audiotaped account of the experience.

2. **Reflecting**   In this step you record and assess your own reactions to the experience. You may respond to particular questions or you may use a less structured format.

3. **Identifying Relevant Theory and Knowledge**   Here you seek out or recall ideas that can help you make sense of the experience in a variety of ways.

4. **Examining Dissonance**   Now you review all the ways you have looked at the experience to see whether there are any points of conflict. These conflicts may be between or among competing theories; between what the theory says should happen and what actually did; between what you believe and what the agency seems to value; or between any two or more aspects of the experience. Sometimes this dissonance is resolvable, and sometimes it is not.

5. **Articulating Learning**   Here you look back over your writing and thinking and write down the major things you have learned from thinking about this experience.

TABLE 1.1
Keeping a Journal: Reflecting on Events

Actions	Thoughts	Feelings
I am sitting in the lounge with several residents. John walks in and sits down. There are several chairs available, but he sits right in front of me.	This guy is always looking for trouble. What is he doing?	Nervous. Uncomfortable.
I say "hello." He nods. I continue my conversation with the residents.		
John squirms around in his chair several times. Finally, I notice the outline of a pack of cigarettes in his pants pocket (a clear violation of house rules).	What is his problem? Damn. He has cigarettes and I'm supposed to take them away and take points. He has a terrible temper. He set this whole thing up.	Annoyed. Anxious. Angry.
When I look up, he is looking right at me.	I have to do something now. He knows that I saw them.	Embarrassed. More anxious.
I say, "What've you got in your pocket there, John?"		
John: "Where?"	Here we go.	
Me: "Right there."		
John: "Nothing! What are you talking about?"	He's not going to make this easy. I'm trying to be nice.	Nervous, angry.
Me: "The cigarettes. You obviously wanted me to see them."		
John: "I did not! So what are you going to do about it anyway?"	If I punish him now, he's going to do something worse.	Confused. Uncertain.
Me: "What do you think I should do?"	I'm stalling and he knows it.	Stupid.
John: "I think you should leave me alone." His face is getting red.	I'm tired of this nonsense.	Angry. Resentful.
Me: "If you wanted that, you shouldn't have come in here. You could have just gone outside and smoked, you know."	I can't believe I just suggested he break a rule—I just wouldn't have had to deal with him if I hadn't seen him.	Upset at myself.
John: "Go take a flying leap [expletives deleted]!" He jumps to his feet.	Uh oh! Are there any other staff around? I have to calm him down. The other kids are watching me.	Scared. Self-conscious. Alert.
Me: "Look, if you just give me the cigarettes, I won't report this."	Maybe this will work.	Hopeful.

(continued)

TABLE 1.1 *(continued)*		
Actions	**Thoughts**	**Feelings**
John: "They're mine. No one takes my property!" He is clenching his fists.	He's not going to get physical over this. Is he?	Frightened.
Me: "You're not allowed to have them here and you know it. I should take points away."		
John: "Do I get them back?"	I can't give them back.	Confused. Desperate.
Me: "I don't know. I'll think about it."		
John: "All right, but only because I like you."		
Me: "Thanks."	Thank God. I wonder if I did the right thing, though.	Relieved. Embarrassed. Angry.

From Sweitzer, in Harris & Maloney (Eds.) *Human Services: Contemporary Issues and Trends,* 2/e. Published by Allyn & Bacon, Boston, MA. Copyright © 1999 by Pearson Education. Reprinted by permission of the publisher.

6. **Developing a Plan**   This is where you consider the next steps in your learning and your work. You may identify areas you need to know more about and places to pursue that knowledge. In addition, you may identify new goals or approaches you plan to use in your work. Taking these next steps can be another new experience, and the cycle can begin again.

Process Recording   Process recording is a widely used reflective technique (Sheafor & Horejsi, 2003) that can be used on its own or in combination with other techniques such as the Integrative Processing Model. It is a way of writing about your work that was developed to use in recording interactions with clients, but with just a little modification it could be used in settings where you are not doing individual or group work as well. There are several variations on process recording, but Sheafor and Horejsi (2003) suggest the following possible components:

- Names and locations of those in attendance
- Date, location, and length of session
- Purpose
- Goals
- Description of the session, including topics discussed, new information learned, significant exchanges, and the overall mood
- Techniques, skills, and roles used
- Assessment of client's major concerns or problems
- Plan for next meeting

SOAP Notes SOAP notes are another common reflective technique (Baird, 2002). This approach asks students to divide their reflections into categories: Subjective, Objective, Assessment, and Plan. These terms are fairly self-explanatory, although Baird (2002) notes that there can be some difficulty deciding exactly in which category a particular observation belongs.

DART Notes Baird (2002) offers an alternative to SOAP notes called DART. This system begins with Description, in which you describe the individuals and situation. This step includes the *who, what,* and *where,* but not the *how* or *why.* In the Assessment step, you use various theories to make sense of the experience and speculate about why events unfolded as they did. In the Response section you record as completely and accurately as possible what you did during the session or incident. And finally, you offer thoughts about the Treatment plan.

You can see from even a casual reading that there is some overlap among these techniques. This should come as no surprise, given that they are all pursuing the same general goal. We believe that any or all of them can be used productively. We also believe that each takes time and practice to learn to use them well. Choose your reflective techniques with care, in consultation with your site supervisor and campus instructor; pursue the resources listed at the end of this chapter to learn more about them; and make sure you are able to get feedback and guidance as you learn to use them.

PORTFOLIOS

Another use for your journal is as a portfolio of your experience. You can use a portfolio as a personal record, but many academic programs require them. Portfolios are also valuable assets at an interview for a job, for graduate school, or even for another internship. If you are interested in using your journal in this way, start planning for and discussing that project with your instructor. There may be items you will want to remove, such as highly sensitive comments about a client or agency. There may also be items you want to include, such as samples of your work, photographs, brochures, and so on. Your instructor can help you think about this now, so that you know what sorts of items to seek out and save, and so that you avoid violating confidentiality.

The Seminar

Most of you will be meeting with an instructor and other interns on campus during the semester. We refer to these meetings as "seminars." The word *seminar* comes from the Italian *seminare,* which means "to sow or seed." The class sessions are a medium that is most helpful in the integration of intellectual and affective learning, encouraging new understanding and creative responses, and strengthening the effectiveness of interpersonal relationships (Williams, 1975).

A seminar may be a bit different from other classes you have taken. For example, one basic assumption of a seminar is that each person has something to contribute (Royse, Dhooper, & Rompf, 2003), unlike many classroom experiences where the assumption often is that only the teacher has something to contribute. A seminar class is one in which an exchange of ideas takes place, where information is shared, and mutual problems are discussed. It is also a forum for problem-centered learning. If everyone has something to contribute, everyone shares the responsibility for the success of the experience. You have additional responsibilities, then, but also additional benefits.

The importance of the seminar to the quality of the internship cannot be overstated. A workshop given at the National Society for Experiential Education described the seminar as "a keystone to learning" (Hesser & King, 1995), because the quality of learning is enhanced when interns come together as a community of learners.

An effective seminar affords opportunities for reflective dialogue, support, the development of important relationships, and a variety of new learning experiences. Other benefits of a seminar include enhancement of individual learning, integration of cognition and affect, strengthening of humanistic values, and increased interpersonal effectiveness (King, 1995). The seminar is also an opportunity to learn how other agencies and placement sites operate and how they approach common challenges and problems. You may engage in many different activities as part of the seminar, including lectures, discussions, reflective assignments, journal writing, student presentations, and support groups (which may meet in or outside of class).

Earlier we mentioned Web-based systems such as Blackboard and WebCT as means of sending and receiving journals. These systems are also valuable for conducting seminars for interns whose placements or program makes meeting in one location impossible. Discussions can be had in real time, which means they are live, as in a chat room; or you can use a bulletin board format (Landsberger, 2002), which allows students and instructors to read and leave comments when it is convenient for them. Of course, both approaches can be used.

Even if you are not working at a distance, you and your instructor may want to consider using some sort of electronic means to send journal entries and comments to each other, or even to the class as a whole. One of us used to have interns hand in their journals each week, and would return them a week later in class. That approach worked well for communicating with individuals, but not for planning seminar sessions. The information about concerns and issues was a week old, and often the concerns had shifted, sometimes dramatically. Now, students submit their journals electronically a day or two before the seminar class.

Regardless of the system that is used, the main purposes of the seminar are skill development, support, and feedback. Skill development needs will vary quite a bit from situation to situation, but support is a common thread in all internship seminars. It is important that interns have a place where they can talk about their experiences, their feelings and reactions, and their struggles and achievements. Although your friends and family can do some of this for you, it is often helpful to have this exchange with others who are undergoing a similar experience. Support groups exist for almost every purpose; perhaps you have participated in some. While the seminar is not a support group in the formal sense, one of its principal benefits is the quality of connections that you develop with your peers. Through these relationships, you give and receive support. In fact, you receive a double benefit: not only do you give and receive support, but you also become more skilled at each of these functions.

CREATING A COMMUNITY OF SUPPORT

Effective support is both very simple and quite difficult, especially when someone is sharing a problem. We have found two principal barriers to creating a community of support in a seminar: a tendency to rely on the instructor and the common cry, "I don't know what to say." A pattern we have observed many times in seminars, and been told about many more, is one where a student recounts an event or poses a problem and the instructor responds. Then the next student recounts, the instructor responds, and so it continues. Even when the instructor pauses for or invites student participation, sometimes it does not come. Students can learn a lot from observing the interactions between the instructor and their peers, but it is not a particularly effective way to build their own skills and sensibilities as givers of support. When we have asked our students about this phenomenon, they have responded by saying they did not think they could be as effective or insightful as the instructor. That is rarely true, but even if it is, the point is to learn and grow. After all, chances are the instructor has been at it for a long time! Another response we get is that they don't know what to say. That issue requires a more detailed treatment.

Theories about and guidelines for effective communication abound in the helping professions. You may very well have read some of them. We offer here, as one way to think about responding, categories of responses suggested by David Johnson in his extraordinary book, *Reaching Out* (Johnson, 1999). Imagine that you or a peer has just told a story about something at the internship. A client has relapsed, a coworker or supervisor has been very harsh in criticizing you, a community meeting turned into a shouting match, or whatever other event you want to imagine.

One category of response is *Advising and Evaluating*. If you have taken a helping skills class, you probably know that giving advice is often not the best approach, but when a friend or classmate is struggling, it can be awfully tempting to offer your heartfelt suggestions. And it can be a great relief as well. But it is not always helpful, and it certainly does not empower your peer to confront and resolve the situation—not to mention the next one! We have found that it is usually best to hold your advice until and unless it is requested.

Another category is *Analyzing and Interpreting*. Here, the listener uses theory to interpret what has happened. It may be an opinion about the underlying psychological dynamics of the people involved, a sociological analysis, or any number of other interpretations. These thoughts can be interesting, and contagious as other students move into the intellectual realm they know so well. But they can also be very distracting, especially if that is not the direction in which the person telling the story wants to go.

Yet another way to respond is *Questioning and Probing*. There is a difference here between asking clarifying questions, so you are sure you understand, and asking questions that take the conversation in a different direction. So, if a student tells a story about a client, and someone asks, "Is this the same client you spoke about last week?" that is clarifying, but "Who referred this client to your agency?" is not.

Reassuring and Supporting is another category. The term *support* in this context may be confusing, because we are talking about building support. But as Johnson uses the terms, this way of responding has as its goal to calm the person and reduce the intensity of feelings. Sometimes that is fine, but sometimes it sends the message, "Don't feel bad." And sometimes what the person needs is just to experience their feelings, at least for a few minutes. Sometime the listener will do what we call a "me too," relating a similar incident that happened. It may be helpful for someone to hear whether others have had a similar problem and how they handled it, but let the request come from that person.

The final category is *Paraphrasing and Understanding*, meaning that the goal of these responses is to make sure that the person is able to explain and elaborate fully the actions, thoughts, and feelings that occurred. There are many specific techniques for achieving this goal, and there are resources suggested for you at the end of the chapter.

OK, so now you are really confused and intimidated. Which one should you use? For better of for worse there is no pat formula, but we do have some suggestions that we think work most of the time.

The most important component of support is listening, so start there. It sounds simple, but think about how rare it is that someone really wants to listen to you, especially when you are struggling. Attention wanders, small talk intrudes, unwanted advice is given. It is a wonderful experience just to have someone listen quietly, attentively, and empathically to whatever you want to say. It is also a wonderful gift to give another person. As one student put it:

> Knowing that others are feeling similar feelings doesn't make those feelings go away, but it does make me feel better about having them. I feel more comfortable now opening up to my classmates and feel better equipped in encouraging them.

Listen carefully, and listen actively. Use skills such as paraphrasing to be sure that you understand the person is trying to say. You will often find that these techniques also help the person to elaborate even more.

Once you have listened, and listened carefully, we suggest that you ask the person what she or he needs at the moment. Advice, analysis, problem solving, or reassurance can be wonderful if they are wanted. Or it may be enough just to be listened to.

It may be, too, that the person would like some feedback, and this is another skill. You have almost certainly studied the principles of effective feedback, but they bear repeating (Porter, 1982). Effective feedback is specific and concrete, as opposed to vague and general; it should refer to very specific aspects of the situation being discussed. It is descriptive rather than interpretive. You can tell people how you feel about what they did or did not do, but it is not your job to tell them why. It is usually best delivered using an "I" statement, rather than a "you" statement. "When you described the situation I had a hard time understanding what was happening" is better than "Your story was really confusing." If you have feelings about what was said, state them directly, and attach them to a specific statement or portion of the story. And finally, feedback should always be checked with the receiver, to see whether you have been understood and whether you have understood the other person.

In trying to build a community of support, it is also important to remember that the seminar class is a group, and it has developmental stages and group dynamics like any other group. It is not hard to describe the atmosphere that you would want in a group where people are sharing ideas, joys, and fears. But the atmosphere of trust, openness, safety, honesty, and feedback that characterize successful groups (Baird, 2000) does not just happen. It happens in stages, over time.

Please remember that this is not a therapy group. You are not therapists (nor are your instructors in therapists' roles, although they may have those skills). There may be times when certain individuals encounter a challenge at the internship or an experience that touches something inside them that needs the attention of a counselor or therapist. Your instructor can help you recognize those instances and locate appropriate resources to deal with them.

Finally, if you are seeking to build a community of support, then leave some time in the seminar to reflect on how that endeavor is going. Early in the seminar, you may want to discuss the modes of responding, the principles of effective feedback, and the overall goal of learning to function as a supportive group. As the semester progresses, take time to celebrate your successes and growth in this endeavor, to discuss and try to solve problems, and to give one another feedback about the achievement of the groups goals.

OVERVIEW OF THE TEXT

Chapter Organization

This book is organized into four sections. In Section One, we present the conceptual framework that underlies the book. Following this chapter, Chapter 2 introduces you to the first big idea underlying this book—stages of an internship. Chapters 3 and 4 are reflective in nature and focus on self-understanding (the second big idea), and we ask you to look at and think about yourselves through a variety of lenses. Chapter 3 is especially theory intensive. Many different approaches to self-understanding are discussed, and we have tried to be as user-friendly as possible in discussing the theories. Although it is perfectly possible to use and integrate all of the theories, we anticipate that some instructors and/or students may want to choose which of the theories and approaches they will emphasize, especially if some of the theories and ideas are new to their programs.

In Section Two, we deal with the issues and concerns associated with getting started in an internship. We call this the *Anticipation* stage. Some of our colleagues and students have told us that they prefer to deal with some of these issues and concerns before the actual placement, and this section could also be used in that way. Chapter 5 discusses the Anticipation stage itself, and Chapters 6 and 7 help you become oriented to clients, coworkers, supervisors, and the agency itself. These two chapters could easily be read out of order. Chapter 8 helps you look at some initial concerns from an organizational perspective, and Chapter 9 asks you to consider the community context of your work.

In Section Three, we look at the challenges that await you after the initial concerns have been resolved. In Chapter 10, we help you take stock of where you are and where you want to be. We also try to help you anticipate some of the more common problems that interns encounter at this point. The final portion of Chapter 10 deals with the *Disillusionment stage*, which can be a disheartening stage for students, faculty, and site supervisors alike, but which we view as a normal and necessary part of the experience. In Chapter 11, we give you the tools to move from disillusionment to *Confrontation* as you attempt to identify and resolve issues that are standing in the way of your continued progress. We emphasize that this process is a learning experience in and of itself as well as one that paves the way for future learning. There is a problem-solving model presented in Chapter 11, although a different model could also be used if there is one you or your instructor prefers, or to which you have already been exposed.

In the final section of the book, Section Four, we examine issues and concerns that are common in the latter stages of the experience: *Competence* and *Culmination.* Chapter 12 deals with several professional issues that await you at this point in the internship. Chapter 13 deals with legal and ethical issues. Of course, ethical issues can arise throughout the internship, but it has been our experience that interns often do not *notice* them, regardless of whether and how often they are covered in class, until some other concerns have been resolved. We suggest some common ethical issues that arise in an internship and also provide a decision-making model. Students who have had a course on ethical issues, or who have encountered this theme in many of their courses, may already have a model for thinking about and resolving ethical dilemmas. However, the issues raised here are probably new to you, or you may be encountering them for the first time in a professional context. The final chapter of the book, Chapter 14, attempts to help you end the internship in a productive way. It covers ending well with clients, coworkers, and supervisors, and there is a final reflection on the experience.

Chapter Exercises

At the end of each chapter, we offer you several ways to extend and enhance your learning. There is a list of additional resources because there may be areas you want to explore further. There are also several exercises, and these fall into two categories. In the section "For Personal Reflection" we pose questions for you to think about and perhaps to write about in your journal. You should select, in cooperation with your instructor, those questions that seem most relevant and meaningful to you. In the section "Springboards for Discussion" we offer questions and exercises that could (in most cases) be undertaken individually but that could also be topics for group discussion or skill building.

We wish you good luck in your journey through the internship, practicum, co-op, service-learning, or community service experience. Now take some time to reflect, before we move on to the major ideas underlying this book.

For Further Reflection

FOR PERSONAL REFLECTION:
SELECT THOSE QUESTIONS MOST MEANINGFUL TO YOU

1. The beginning of your internship or service-learning experience is a good time to review your academic program's expectations of you, your supervisor, and your instructor during the internship. This is particularly important in terms of knowing what you can expect from others and what others, including the staff at the

placement site, might expect from you. Take time now to locate any written documents from your program that specify these responsibilities. Make copies and keep them with your other internship paperwork.

2. As you begin your journal, you need to decide whether it will be handwritten, tape-recorded, typed, or computer generated. Then you will need to make sure you have the materials you need. It is also time to clarify the larger purposes of your journal, who will have access to it, and how you want to set it up to meet those needs.

3. Review the kinds of journals and reflective techniques discussed in this chapter. Which ones seem interesting and useful to you? Are any of them required for you?

4. Consider using your journal as a portfolio of your internship. What sorts of things might you include? If this is not appropriate in your case, you can always keep a portfolio that is separate from your journal.

SPRINGBOARDS FOR DISCUSSION

1. Internship students often use a language of their own. Your supervisor or coworkers may appear puzzled when you use certain terms, even though they are commonly understood on your campus. We call this language *fieldspeak*. There is also *agency-speak*, which you may adopt without even thinking about it after a few days at the placement, but it will puzzle your seminar classmates or even your instructor. Review your program's definitions of terms and compare them to the ones in this chapter. Be ready to explain them to people at your placement site. In class, share the important terms and slang that make up your field site's agencyspeak.

2. Seminar class can become an important part of the internship experience. Now is a good time to think about what you want from that class. For example, what are your major objectives for the time spent in seminar? What works best for you as a way of teaching and learning? What role do you see yourself having in class? In class, share these ideas with your peers and instructor. Make a list of the major goals and objectives and learning preferences and discuss ways you can help one another.

3. Choose a real or hypothetical experience that happened at an internship, and use it to illustrate the categories of responding. Have one intern tell the story and then have others take turns responding, each in a different category of response. One gives advice, one analyzes, and so on. Then the person who told the story describes how each response felt. Afterward, as a group, talk about which modes of responding were easier to do and which ones may need practice.

4. Devote a class, or a portion of the class, to practicing effective feedback. Again, you may use a real or a scripted situation. Have each person practice giving feedback, and have others use a checklist or rating scale to point out strengths and areas for development. This exercise could also be done in triads or small groups.

For Further Exploration

Baird, B. N. (2002). *The internship, practicum and field placement handbook: A guide for the helping professions* (3rd ed.). Upper Saddle River, NJ: Prentice Hall.

Thorough discussions of a variety of writing and recording techniques.

Benjamin, A. (1987). *The helping interview.* Boston: Houghton Mifflin.

Especially helpful section on different ways of responding.

Burger, W. R., Youkeles, M., Malamet, F. B., & Smith, F. (1999). *The human services professions: A careers sourcebook.* Belmont, CA: Wadsworth.

An excellent discussion and taxonomy of a variety of different careers, with information on training and employment opportunities for each one.

Eyler, J., & Giles, D. (1999). *Where's the learning in service learning?* San Francisco: Jossey-Bass.

A thorough discussion of service learning concepts and the results of the first large-scale, systematic study of the impact of service on learning.

Johnson, D. W. (1999). *Reaching out: Interpersonal effectiveness and self-actualization* (7th ed.). Boston: Allyn & Bacon.

An excellent discussion of interpersonal communication skills. Loaded with exercises, charts, and other learning aids.

Kiser, P. M. (1998). The integrative processing model: A framework for learning in the field experience. *Human Service Education, 18*(1), 3-13.

Kiser, P. M. (2000). *Getting the most from your human service internship: Learning from experience.* Belmont, CA: Wadsworth.

Both of these sources have excellent, clear descriptions of Kiser's model.

Russo, F. X., & Willis, G. (1986). *Human services in America.* Upper Saddle River, NJ: Prentice Hall.

A good introduction to the six major categories of human service delivery systems as well as an introduction to public policy in human services.

References

Baird, B. N. (2002). *The internship, practicum and field placement handbook: A guide for the helping professions* (3rd ed.). Upper Saddle River, NJ: Prentice Hall.

Bringle, R. G., & Hatcher, J. A. (1999). *Introductory service-learning tool kit.* Providence, RI: Campus Compact.

Chickering, A. W. (1977). *Experience and learning: An introduction to experiential learning.* Rochelle, NY: Change Magazine Press.

Dewey, J. (1933). *How we think.* Lexington, MA: D. C. Heath.

Dewey, J. (1940). *Education today.* New York: Greenwood Press.

Dewey, J. (1944). *Democracy and education.* New York: Macmillan. (Original work published 1916).

Dewey, J. (1963). *Experience and education.* New York: Macmillan. (Original work published 1938).

Eyler, J., & Giles, D. (1999). *Where's the learning in service learning?* San Francisco: Jossey-Bass.

Garvin, D. A. (1991). Barriers and gateways to learning. In C. R. Christensen, D. A. Garvin, & A. Sweet (Eds.), *Education for judgment* (pp. 3–13). Boston: Harvard Business School Press.

Giles, D. (2002, October). *Assessing service learning.* Address to the National Organization for Human Service Education, Providence, RI.

Hesser, G., & King, M. A. (1995). *Internship seminar: A keystone to learning.* Workshop presented at the annual meeting of the National Society for Experiential Education, New Orleans.

Inkster, R., & Ross, R. (1998, Summer). Monitoring and supervising the internship. *NSEE Quarterly,* 10–11; 23–26.

Johnson, D. W. (1999). *Reaching out: Interpersonal effectiveness and self-actualization* (7th ed.). Boston: Allyn & Bacon.

King, M. A. (1995). *Toward rewarding field experiences: A guiding framework.* Unpublished manuscript, Fitchburg State College.

Kiser, P. M. (1998). The integrative processing model: A framework for learning in the field experience. *Human Service Education, 18*(1), 3–13.

Kiser, P. M. (2000). *Getting the most from your human service internship: Learning from experience.* Belmont, CA: Wadsworth.

Kolb, D. A. (1984). *Experiential learning: Experience as the source of learning and development.* Upper Saddle River, NJ: Prentice Hall.

Kolb, D. A. (1985). *Learning style inventory.* Boston: McBer.

Kolb, D. A., & Fry, R. (1975). Toward an applied theory of experiential learning. In C. Cooper (Ed.), *Theories of group process* (pp. 33–57). New York: Wiley.

Lacoursiere, R. (1980). *The life cycle of groups: Group developmental stage theory.* New York: Human Sciences Press.

Landsberger, J. (2002). Integrating a web-based bulletin board into your class: A guide for faculty. *Tech Trends, 45*(5), 50–53.

National Crime Prevention Council. (1988). *Reaching out: School-based community service programs.* Washington, DC.

National Society for Experiential Education Foundation Document Committee. (1998). Foundations of experiential education, December 1997. *NSEE Quarterly, 23*(3), 1;18–22.

Porter, L. (1982). Giving and receiving feedback: It will never be easy but it can be better. *Reading book for human relations training* (7th ed.). Arlington, VA: National Training Laboratories.

Royse, D., Dhooper, S. S., & Rompf, E. L. (2003). *Field instruction: A guide for social work students* (4th ed.). Boston: Allyn & Bacon.

Schutz, W. (1967). *Joy.* New York: Grove Press.

Sgroi, C. A., & Ryniker, M. (2002). Preparing for the real world: A prelude to a fieldwork experience. *Journal of Criminal Justice Education, 13*(1), 187–200.

Sheafor, B. W., & Horejsi, C. R. (2003). *Techniques and guidelines for social work practice.* Boston: Allyn & Bacon.

Simon, E. (1989). Field practice survey results. In C. Tower (Ed.), Fieldwork in human services: Council for Standards in Human Service Eduction (Monograph #6).

Sweitzer, H. F., & King, M. A. (1994). Stages of an internship: An organizing framework. *Human Service Education, 14*(1), 25–38.

Van Grinsen, M. T. (1995, June). Company college program receives high marks. *Best's Review,* 69–70.

Weinstein, G. (1981). Self-science education. In J. Fried (Ed.), *New directions for student services: Education for student development* (pp. 73–78). San Francisco: Jossey-Bass.

Weinstein, G., Hardin, J., & Weinstein, M. (1975). *Education of the self: A trainer's manual.* Amherst, MA: Mandella.

Williams, J. K. M. (1975). The practice seminar in social work education. In J. K. M. Williams (Ed.), *The dynamics of field instruction* (pp. 94–101). New York: Council for Social Work Education.

CHAPTER *2*

Framing the Experience: The Developmental Stages of an Internship

Internship is like a diamond, in that it is multifaceted; it is also like a roller coaster with its highs and lows.

Allowing the stages to happen allows the intern to learn and have positive learning experiences.

STUDENT REFLECTIONS

Each intern's experience is unique, and yours will be, too. You may have an experience that's different from those of other interns at the same placement or from any other previous field experiences you have had. Placement sites differ, too; you may be in a seminar with peers who are doing very different work, with very different groups of people, than you are. We continue to be amazed and enriched by the diversity of experiences that interns have; it is one of the factors that makes working with interns gratifying, even after many years. Over time, though, we have noticed some similarities that cut across these various experiences. Some of the concerns and challenges that most interns face seem to occur in a predictable order. Our experience, plus our study of other stage theories, has yielded our own theory of internship stages (Lacoursiere, 1980; Sweitzer & King, 1994, 1995).

THE DEVELOPMENTAL STAGE MODEL

The focus (of the week) has been for me to normalize my feelings and allow the process to happen.

STUDENT REFLECTION

26

There are five developmental stages of an internship: Anticipation, Disillusionment, Confrontation, Competence, and Culmination (Table 2.1). The stages are not completely separate or discrete; concerns from earlier or subsequent stages can often be noticed in any of the stages. However, certain concerns and issues are apt to be particularly prominent during each stage, along with associated feelings or affect.

Each stage has its own obstacles and its own opportunities. You will have certain concerns during each stage, and to some extent, you must resolve those concerns in order to move forward and continue learning and growing. The process of resolving the concerns is also a learning experience in and of itself. In each stage, there are important tasks that will help you address the concerns; however, we cannot predict how quickly you will move through the stages.

An important distinction for understanding the stages is that between morale and task accomplishment (Lacoursiere, 1980). The term *morale* refers to the interpersonal and intrapersonal tone of your experience at the agency. High morale is characterized by positive feelings about yourself, your work, and the agency. The tone is one of hope, optimism, and enthusiasm, and there is movement toward goals, even in the face of obstacles. As much as you would probably like to have high morale at all times, that is not usually what happens. The good news is that morale can often be recovered when it drops, and there is great learning in the process that only occurs if you fully experience both the drop and the recovery.

The term *task accomplishment* refers not so much to the specific tasks assigned by the placement site, but to the attitudes, skills, and knowledge that you hope to acquire. Of course, there may be considerable overlap between the tasks you are given and your learning goals. Here again, you might hope that the growth of this dimension would follow a steady, linear, upward path, but both our experience and some research suggest that this is not the case (Blake & Peterman, 1985). There will be periods when you are learning and growing at an incredible pace. There will also be periods where you feel stuck, and you may be tempted to think you aren't ever going to get where you want to go or that you aren't learning anything. You are always learning, though, or at least the opportunity is always there; paying attention to what you are learning rather than dwelling on what you are not will help you get back on track.

As we mentioned earlier, your rate of progress through the stages is affected by many factors, including the number of hours spent at the agency; previous internships or field experiences; your personality; the personal issues and levels of support you bring into the experience; the style of supervision you receive; and the nature of the work you do, including the emotional issues stimulated by a particular client population.

Stage 1: Anticipation

Before I read this chapter, I thought that I was the only one experiencing these anxieties.

This stage reminded me of my first year in high school and college. I just wanted to be accepted and didn't know how to do that. Although I am starting to gain a sense of what is expected from me in the internship, I am still wondering what staff members and clients think of me. Do they think I am stupid, lazy, ignorant, etc.?

STUDENT REFLECTIONS

As you look forward to and begin your internship, there is usually a lot to be excited about. Students often eagerly wait for an internship for several semesters; it is your best chance to actually get out there, do what you have wanted to do, and make a contribution to others. For most interns, however, along with the eagerness and hope, there is inevitably some anxiety. It may not be very visible, even to you, but there are enough unknowns in the experience to cause some concern and anxiety in anyone.

For interns, this anxiety generates the first set of concerns, which generally center on the self, the supervisor, clients, and coworkers. We often refer to this as the "What if…?" stage because interns wonder about things like: What if I can't handle it? What if they won't listen to me? What if they don't like me? or, What if my supervisor thinks I know more than I really do? You will probably be concerned about what you will get from the experience and what it will really be like to work at this site. Many interns wonder whether they can "really do this" and what will be expected of them.

Some interns report fears that they are not competent, that they have gotten this far only by great luck, and that in their internship they will surely be found out. You may also wonder about your role; you are not in a student role while at the placement, but you are not a full-fledged staff member either. Depending on your personal situation, you may also be concerned about your family and the effect that your participation in such a demanding experience will have on them.

You are going to interact with a number of people during your internship, and it is natural to wonder what to expect from them and whether they will accept you in your new role. You may, for example, be unsure of the role and responsibilities of the site supervisor; you will be unusual if you don't wonder what your supervisor will think of you and whether he or she will care about you. Those working directly with clients inevitably wonder about how they will be perceived and accepted by clients and just what kinds of behaviors and problems clients are going to exhibit. Finally, most interns are concerned about the reception and treatment they will receive from agency staff members. You may also wonder how you are going to manage the other responsibilities in your life and who is going to be there to support you.

The level of task accomplishment at this time is often relatively low, meaning that you may not be learning the specific things you went there to learn, and that can be frustrating. What is most important at this stage, however, is that you learn to define your goals clearly and specifically and begin considering what skills you will need to reach them. You must also develop a realistic set of expectations for the experience. Since you have not yet actually experienced the internship, but have probably thought and maybe heard a lot about the agency where you will be working, it is inevitable that you will make assumptions, correctly or incorrectly, about many aspects of the internship (Nesbitt, 1993). Some of these assumptions come from stereotypical portrayals in the media of certain client groups (such as the mentally ill) or agencies (such as detention centers); others may come from your own experience with certain issues or problems. As much as possible, these assumptions and expectations need to be made explicit and then examined and critiqued. Finally, you need to work on being accepted by and developing good relationships with your supervisor, coworkers, and clients.

Stage 2: Disillusionment

Issues concerning doubt and confrontation are conflicting for me. I cannot imagine having my expectations crushed...
   STUDENT REFLECTION

Sooner or later, you will probably reach a time when you are not as certain or as positive about your internship as you would like to be. You may find that you are having trouble getting up and going to the internship or that you are mumbling under your breath or complaining to friends. It is an unusual but not unheard of intern who does *not* experience some kind of disappointment at some point. Consider these two very difference perspectives on this experience:

The most critical learning incident of the week was by far the decision to review the stages...most particularly stage 2...dissatisfaction...it was right on. It was important because it reassured me that my feelings of confusion, frustration and the emotions were normal...essential in keeping the intern in touch with reality during [the] internship. It allows me to relax and allow the experience to be true and legitimate. Previously I had thought of "where am I going with this...how will this work?"

This stage was interesting to me because I could not relate [to it] — I love getting up in the morning for this internship and going on with more education. I have never, so far, complained about it, even though it is still so early in the semester. I truly hope that I skip this stage. I find myself wanting to observe and participate in everything.

   STUDENT REFLECTIONS

When the shift occurs, as it so often does, one reason is that there is almost always a difference between what you anticipated about your internship and what you really experience.[1] The size of this gap, and hence the dip in your morale, will depend on how successfully you accomplished the tasks of the Anticipation stage, but it cannot be avoided altogether. If the concerns of the Anticipation stage have been addressed, you will be less likely to encounter a wildly different reality from what you expected, but there will be some discrepancies, and some of them will be troubling. Furthermore, issues will arise that you simply never considered. For some of you, the change will be subtle and barely noticeable; for others, it will be profound and overwhelming.

---

[1] In some cases, the intern's primary focus is not the work of the internship but the location. For example, some students use an internship in part as a way to be in a major city, like New York or Washington, D.C., or even to go abroad. In these cases, disillusionment, if it occurs, is likely to focus on some of the harsh realities of life in a new place. Of course, some interns are focused equally on the work and the location. For them, there may be two possible sources of disillusionment.

TABLE 2.1
Developmental Stages of an Internship

Stage	Associated Concerns	Response Strategies

1. Anticipation	Positive expectations
	Anxieties
	  Self
	    Role
	    Appropriate disclosure
	    Self in authority role
	  Supervisor
	    Supervisory style
	    Expectations of disclosure
	    Perception and acceptance
	    Assessment
	  Coworkers
	    Organizational structure
	    Standards of behavior
	    Acceptance
	  Field Site
	    Philosophy, norms, values
	    Workload
	    Hiring potential
	  Clients
	    Acceptance and perception
	    Needs and presenting problems
	  Life Context
	    Responsibilities
	    Support system
		Realistic, clear, specific goals
		Clarify and assess expectations
		Make an informed commitment

2. Disillusionment	Unexpected emotions
	  Frustration
	  Anger
	  Confusion
	  Panic
	Adequacy of skills
	Breadth of demands
	Relationship with clients
	Operating values of organization
	Disappointment with
	  supervisor/coworkers
		Acknowledge gap between
		  expectations and reality
		Normalize feelings and behaviors
		Acknowledge and clarify specific issues
		Acknowledge and clarify feelings

3. Confrontation	Achieve independence Gain confidence Experience effectiveness Changes in opportunities Interpersonal issues Intrapersonal blocks	Reassess goals and expectations Reassess support systems Develop specific strategies

TABLE 2.1 *(continued)*		
Stage	**Associated Concerns**	**Response Strategies**
4. Competence	High accomplishment Investment in work Quality supervision Ethical issues Worthwhile tasks Home/self/career issues	Share concerns openly Develop coping strategies

5. Culmination	Termination with clients
	Case management issues
	Redefine relationships with
	  Supervisor
	  Coworkers
	  Faculty
	  Peers
	Ending studies
	Post-internship plans
		Identify feelings
		Recognize unfinished business
		Meet with supervisor
		Gather with colleagues
		Write final reflection

If the Anticipation stage was the "What if . . . ?" stage, then the Disillusionment stage is the "What's wrong?" stage. Concerns in this stage center on many of the same areas as earlier in the placement: clients, supervisor, the agency, the "system," or yourself. However, feelings associated with these concerns often include frustration, anger, sadness, disappointment, and discouragement. You may find yourself directing any or all of these feelings toward the site supervisor, your instructor, clients, coworkers, or even yourself.

This stage is the onset of what we refer to as a *crisis of growth.* It is possible to become stuck in this stage, and that can have unfortunate consequences: at best, learning and growth will be limited; at worst, the placement may have to be renegotiated or even terminated. On the other hand, letting yourself feel the impact of these issues and working them through present tremendous opportunities for personal and professional growth.

Stage 3: Confrontation

I recognized, sensed and felt the great struggle exploring change can initiate within others as well [as me].

Maybe it was the word confrontation that caught my attention. . . . I was still reflecting on a recent confrontational situation at my site. . . . I began to read [about the stage], but after a few pages, I stopped. I had two immediate reactions: one, I wish I had this material a few weeks ago, it would have helped me greatly . . . and two, I would prefer to read [this] text from start to finish . . .

*[even now that my internship has ended] . . . to follow the internship process
"postmortem."*

*Once I confronted these anxieties, though, everything worked out. I am
beginning to confront this situation by making something good of this.*

STUDENT REFLECTIONS

As the saying goes, "The only way around is through." The way to get past the Disillusionment stage is to face and study what is happening to you. Some interns resist acknowledging any problems, even when their level of task accomplishment is dropping. You may fear that any problems must somehow be your fault or that you will be blamed for them. You may think that "really good" interns would never have these problems. Paradoxically, though, it is the failure to acknowledge and discuss problems that can diminish your learning experience, your performance, and your evaluation by supervisors on-site and on campus (Blake & Peterman, 1985).

Moving through this stage often involves taking another look at your expectations, goals, and skills. Although your goals may have seemed reasonable when you set them, experience may have shown that some of them were not realistic, or the opportunities may have changed. This is also a time to reexamine and perhaps take the necessary steps to bolster your support system.

There may be interpersonal issues between you and your clients, supervisor, or coworkers that are getting in the way. You may need some help clarifying these issues and developing a strategy for resolving them. You will need to consider intrapersonal factors, such as mounting personal problems or unexpected crises in your outside life. There may also be aspects of your personal makeup that are contributing to the problems. For example, it may be that your reactions to some typical features of an internship (such as criticism, authority, or speaking in or before a group) reflect patterns evident throughout your life that are being exacerbated by the internship. There are many strategies for dealing with these intrapersonal issues, and we will explore some of them in Section Three of this book.

As the issues raised in the Disillusionment stage are resolved, morale begins to rise, as does task accomplishment. Your task at this stage is to keep working on the issues raised. This is a time when you may be tempted to "freeze the moment" and resist raising any more issues for fear of spoiling the progress you have made. The temptation is normal, but if you give in to it for long, you may find yourself stagnating or even regressing. However, with each new round of confrontation, you will feel more independent, more effective, and more empowered as a learner. You will have a sense of confidence that comes not just from what you have accomplished, and not from denying problems, but from your knowledge that you can grapple with problems effectively.

Stage 4: Competence

*The Competence stage consists of being confident in myself. I am looking
forward to that aspect of my internship. It's not that I have never been
competent in a task before. It is just that the sense of professionalism will be
much greater.*

[The site supervisor] gave me some good guidance and lots of space to create what I wanted. This is usually a good thing for me. To be left alone to do what I wish. I felt intimidated by my audience so I want to be perfect in what I present and not look stupid. So, I have had some difficulty getting it all together.

While I was aware that I had experienced a difficult day [at work], my focus at the internship at night was to be more attentive. It meant putting into perspective all that occurred during the day in order to be effective at night. Balancing the work during the day continues to be the hard work of negotiating the intellect and the affect.

STUDENT REFLECTIONS

As your confidence grows, you will forge ahead into a period of excitement and accomplishment. This is the stage that every intern looks forward to—the reason for the internship. Morale is usually high, as is your sense of investment in your work. Your trust in yourself, your site supervisor, and your coworkers often increases as well. You may find yourself thinking of yourself less as an apprentice and more as a professional. You may even wonder why you are not being paid.

As an emerging professional, you have a solid platform from which to expect, or even demand, more from yourself and your placement. You may find that you want more than you are getting from your assignments, your instructor, or your supervisor. Many interns also report that during this time they are better able to appreciate the ethical issues that arise in their placements and are more willing to confront them. These are all positive developments. If taken too far, though, they can lead to perfectionism. You may begin to apply unreasonable standards to those around you, to yourself, or both. Excellence, not perfection, is your goal in this stage.

Another issue that can arise during this time is the stress of juggling your life outside the internship with your increasing commitment to your work. Although you may feel pulls on your time and loyalty throughout your placement, your earlier anxieties and roadblocks may have demanded too much of your attention to think about these conflicts. Now that these earlier crises are past, conflicts between the internship and home, school, or friends can surface more easily. This can become overwhelming, especially if you strive for perfection rather than excellence in all these arenas.

Stage 5: Culmination

I looked forward to entering the Competence stage, but I am not looking forward to stage 5—when placement ends.

The end of this experience will be sad, . . . everything is ending . . . but the reflections will be great.

STUDENT REFLECTIONS

This stage occurs as your internship approaches its ending date. The end of the internship, coupled with the end of the semester and in some cases with the end of the college experience, can raise some big issues for you. You may experience a variety of feelings as this time approaches. Typically, there is both pride in your achievements and some sadness over the ending of the experience. You may also feel guilty about not having done enough for clients or concern that no one will be as effective with certain clients as you have been (and you may be right about that).

For those of you who are ending your college career, you may be concerned with continuing your education, employment, or economic survival. Relationships with friends, family members, lovers, and spouses that have been organized around your role as a student have to be reorganized. In any case, there are many good-byes to be said. Good-byes are never easy, and for some people, they are very difficult.

Often, interns find ways to avoid facing and expressing these feelings, particularly the negative ones. Avoidance behaviors may include joking, lateness, or absence. Some interns may devalue the experience—they begin saying it hasn't been all that great or find increasing fault with the placement site and/or clients. Many interns find themselves having a variety of feelings and reactions, some of them conflicting and changing by the hour. This can be very confusing and upsetting.

To address the concerns of this stage, you need to focus on your feelings (whatever they may be), have a safe place to express them, and find satisfying ways to say good-bye to clients, staff, supervisors, and in some cases, other interns, both at the site and in an internship seminar on campus.

Of course, if you do not pay attention to the concerns of the Culmination stage, the internship will end just the same. However, you can be left with an empty and unfinished or unfulfilled feeling. In some cases, interns struggling with the Culmination stage actually sabotage their placement by allowing their discomfort about ending to color their perceptions of the entire experience and the affect others' perceptions of their field work.

SUMMARY

I am aware that my stages and concerns are not unique to me.... One strategy was to respond to the concerns of each of the stages The vocabulary [of the stages] gave me a way to express myself . . . to work to normalize my feelings and behaviors...this week I acknowledged feelings, reassessed goals, expectations and support systems, and began to develop strategies. And I shared concerns and started to identify unfinished business. These experiences spurred me to make decisions, [to] choose behaviors I would not have in the past. Euphemistically speaking, it was a busy week. (Two weeks before internship ends.)
STUDENT REFLECTION

Now you have a sense of what is ahead in your internship and in this book. Remember that even though these stages may hold true for many or even most interns, especially when viewed from the outside, both the pace with which you move through the stages and the phenomenological experience of being in them will vary a great deal from individual to individual. As you move through the stages of your internship, you will find that the following chapters in this book explore each stage in more depth and encourage you to remain focused, as well, on the aspects of yourself you will explore in Chapters 3 and 4.

For Further Reflection

I am allowing myself some much needed down time. . . . I received the academic and professional acknowledgment that I have been working toward…truly a bittersweet experience. I never thought that I would regret that it was so short [600 hours]. Now, a time for reflection . . . to catch up on all that I have learned and experienced.
STUDENT REFLECTION

FOR PERSONAL REFLECTION

As you read the description of the stages of an internship, did anything seem remotely familiar? Did the stages remind of you of any other experiences you have had? As you read about the stages, did any one in particular stick in your mind or attract your attention? Why?

SPRINGBOARD FOR DISCUSSION

Think about the issues raised for you by your understanding of the stages, and discuss your thoughts with your peers. Are there issues that the stages do not obviously address? What are your thoughts about knowing about these stages at this point in your placement?

For Further Exploration

Baird, B. N. (2002). *The internship, practicum, and field placement handbook.* Upper Saddle River, NJ: Prentice Hall.

Comprehensive and especially useful to graduate students in the helping professions.

Boylan, J. C., Malley, P. B., & Reilly, E. P. (2001). *Practicum & internship: Textbook and resource guide for counseling and psychotherapy* (3rd ed.). Philadelphia: Bruner-Routledge.

Takes a comprehensive approach to many aspects of graduate counseling internships.

Chiaferi, R., & Griffin, M. (1997). *Developing fieldwork skills: A guide for human services, counseling and social work students.* Pacific Grove, CA: Brooks/Cole.

Offers a developmental framework of stages for the intern-supervisor relationship.

Cochrane, S. F., & Hanley, M. M. (1999). *Learning through field: A developmental approach.* Boston: Allyn & Bacon.

Takes a developmental approach to social work field experiences.

Faiver, C., Eisengart, S., & Colona, R. (2000). *The counselor intern's handbook* (2nd ed.). Belmont, CA: Wadsworth/Thomson Learning.

Takes a focused, pragmatic approach to counseling field experiences.

Gordon, G. R., McBride, R. B., & Hage, H. H. (2001). *Criminal justice internships: Theory into practice* (4th ed.). Cincinnati, OH: Anderson Publishing Co.

Offers the undergraduate criminal justice intern a comprehensive guide to issues specific to a criminal justice internship.

Kiser, P. M. (2000). *Getting the most from your human service internship: Learning from experience*. Belmont, CA: Brooks/Cole.

Offers a framework for processing and integrating previous learning while engaging the field experiences.

Lacoursiere, R. (1980). *The life cycle of groups: Group developmental stage theory.* New York: Human Sciences Press.

Explains Lacoursiere's theory in detail and discusses its application to many kinds of groups.

Schutz, W. (1967). *Joy.* New York: Grove Press.

A group development theory that has had an effect on our view of internships. Talks a great deal about acceptance, inclusion, and control issues.

Sweitzer, H. F., & King, M. A. (1994). Stages of an internship: An organizing framework. *Human Service Education, 14*(1), 25–38.

Gives more details of how Lacoursiere's and Schutz's works informed our thinking about our model of developmental stages of an internship.

Sweitzer, H. F., & King, M. A. (1995). The internship seminar: A developmental approach. *National Society for Experiential Education Quarterly, 21*(1), 1, 22–25.

Discusses our general approach to working with interns in a seminar class from a developmental perspective.

Models Framing Field Experiences

Chiaferi, R., & Griffin, M. (1997). *Developing field work skills.* Pacific Grove, CA: Brooks/Cole.

Cochrane, S. F., & Hanley, M. M. (1999). *Learning through field: A developmental approach.* Boston: Allyn & Bacon.

Gordon, G. R., McBride, R. B., & Hage, H. H. (2001). *Criminal justice internships: Theory into practice* (4th ed.). Cincinnati, OH: Anderson Publishing Co.

Grant, R., & MacCarthy, B. (1990). Emotional stages in the music therapy internship. *Journal of Music Therapy, 27*(3), 102–118.

Grossman, B., Levine-Jordan, N., & Shearer, P. (1991). Working with students' emotional reaction in the field: An educational framework. *Clinical Supervisor 8*, 23–39.

Inkster, R., & Ross, R. (1998, Summer). Monitoring and supervising the internship. *National Society for Experiential Education Quarterly, 23* 4, 10–11, 23–26.

Kerson, T. (1994). Field instruction in social work settings: A framework for teaching. In T. Kerson (Ed.), *Field instruction in social work settings* (pp. 1–32). New York: Haworth Press.

Kiser, P. M. (1998). The integrative processing model: A framework for learning in the field experience. *Human Service Education, 18*(1), 3–13.

Lamb, D., Barker, J., Jennings, M., & Yarris, E. (1982). Passages of an internship in professional psychology. *Professional Psychology 13*, 661–669.

Michelsen, R. (1994). Social work practice with the elderly: A multifaceted placement experience. In T. Kerson (Ed.), *Field instruction in social work settings* (pp. 191–198). New York: Haworth Press.

Rushton, S. P. (2001). Cultural assimilation: A narrative case study of student-teaching in an inner-city school. *Teaching and Teacher Education, 7*, 147–160.

Siporin, M. (1982). The process of field instruction. In B. Sheafor & L. Jenkins (Eds.), *Quality field instruction in social work* (pp. 175–198). New York: Longman.

Skovholt, T. M., & Ronnestad, M. H. (1995). *The evolving professional self: Stages and themes in therapist and counselor development.* New York: Wiley.

Wentz, E. A., & Trapido-Lurie, B. (2001). Structured college internships in geographic education. *Journal of Geography, 100*, 140–144.

References

Blake, B., & Peterman, P. J. (1985). *Social work field instruction: The undergraduate experience.* New York: University Press of America.

Lacoursiere, R. (1980). *The life cycle of groups: Group developmental stage theory.* New York: Human Sciences Press.

Nesbitt, S. (1993). The field experience: Identifying false assumptions. *The LINK (Newsletter of the National Organization for Human Service Education), 14*(3), 1–2.

Sweitzer, H. F., & King, M. A. (1994). Stages of an internship: An organizing framework. *Human Service Education, 14*(1), 25–38.

Sweitzer, H. F., & King, M. A. (1995). The internship seminar: A developmental approach. *National Society for Experiential Education Quarterly, 21*(1), 1, 22–25.

CHAPTER 3

Understanding Yourself

*I have been in my placement for several weeks and have challenged
my own philosophies many times. It frightens me to think that the
very foundation on which I have based my life is being challenged
by clients who I believed were going to be textbook cases. Not that
I assumed that I was entering a vacuum, but I didn't think that
my own beliefs could be shaken in such a short period of time.
Maybe I am making no sense at all. Maybe I am trying to make
too much sense.*

STUDENT REFLECTION

As you prepare to start your placement, what are some of the things you'd like to know? What knowledge, if you had it, would make you feel more prepared?

When over the years we have asked students these questions, we have gotten all kinds of answers, including those focused on clients, coworkers, intervention techniques, community dynamics, and agency rules. Those are all good answers to the questions and important things to know. What we hear less of, and what is just as important, is that you need to know about yourself.

Robert Kegan, a developmental psychologist, believes that the most fundamental human activity is "meaning making" (Kegan, 1982, 1994). Each of us is constantly engaged in a process of trying to understand what is happening to us; we struggle to make sense of the world in which we live, the people in it, and the experiences we have. Perhaps nowhere is this activity more important than in human service work. In your preparation for work as a human service professional, you have been encouraged to consider the complex factors that lead to human problems and their solutions. As soon as you enter your placement, that will stop being an abstract exercise; you won't be thinking about human problems in general anymore, but about the ones that confront you daily. When that happens, you must consider an additional factor in thinking about problems, and that is *you*. You must pay attention not only to the sense you are making of what you encounter, but of *how* you are making sense. Doing so will make you a more effective intern, a more effective practitioner in the future, and more sensitive to your individual journey through the stages of your internship.

Suppose, for example, you are confronted with the following situations at your internship:

- Your supervisor, in correcting your work, makes some insensitive statements.
- One of your coworkers is too aggressive in advancing some ideas.
- One of your clients is backsliding.

Now you must make sense of these events. What factors will you consider? What is the problem and who or what needs to change? The very words we just used to describe the situations imply that the problems, and hence the solutions, lie with other people. However, you are part of the problem too, and not just in what you may have done to cause another person's response. For example, you think your coworker is being too aggressive. Too aggressive for whom? Some people would not find that behavior too aggressive or even describe it as aggressive at all. Why did you? What needs to happen now? Does your coworker need to tone it down, or do you need to learn to be more tolerant?

Self-understanding also plays an important role in helping human service workers fulfill their responsibilities effectively and responsibly. As a human service intern, regardless of your specific duties, you are usually trying to form relationships with clients. In the context of those helping relationships, you attempt to respond, or help clients respond, to a variety of human problems. Although you may never have thought of it this way, in fulfilling these functions you occupy a position of power and influence over clients (Brammer, 1985). Not only do you form opinions and make judgments, but you also make those opinions known to others in counseling sessions, reports, and staff meetings.

Furthermore, you may be asked to recommend for or against various kinds of intervention. Sometimes your placement controls vital resources, in some cases as basic as food and shelter. Increased understanding of the feelings, beliefs, tendencies, and style that make up your system of meaning making will help you be more effective in all your functions. It will also help you be sure you are using your power for the welfare of the clients.

A number of authors have commented on the role of self-understanding in forming effective relationships (Brill & Levine, 2001; Corey & Corey, 2003; Schram & Mandell, 2002). If you can clarify and discuss your feelings and personal patterns, you can serve as examples for clients who are struggling to do the same. Furthermore, if you have confronted and modified aspects of yourself that you didn't like, you are more likely to communicate a belief in the capacity for change (Corey & Corey, 2003). Finally, awareness of ways in which your experiences are similar to those of clients helps establish empathy and trust.

However, you will also deal with many people whose experiences and values are very different from your own. In those cases, self-understanding will help you avoid three major pitfalls: (a) projection, (b) professional myopia, and (c) the tendency to regard difference as deviance (Sweitzer & Jones, 1990). At the heart of combating these tendencies is the ability to see your own reactions and views as only one response among many possibilities. For this reason, understanding some of the sources of your personal characteristics and responses is critical. Understanding how your family, for example, influenced your beliefs and personality, coupled with knowledge of the diversity in family patterns and dynamics, will help you be more objective in your assessment and opinion of others.

Projection, as we are using the term, refers to the tendency to believe that you see in others feelings and beliefs that are actually your own. This tendency can affect your ability to understand, accept, and empathize with another person. If a client brings up an issue with which you are uncomfortable in your own life, you may avoid the subject or decide that the client is being difficult (Corey, Corey, & Callanan, 2003). If you are angry at a client and unaware of that anger, you may decide that the client is angry at the world; if you have trouble with assertiveness, you may react negatively to an assertive client, supervisor, or coworker. Projection can also lead to poor decisions about client needs. You may fall into the trap of inadvertently using clients to meet your own needs (Corey et al., 2003). For example, if you have trouble with assertiveness, you may push clients to be assertive in part because it fulfills your need vicariously. Increased self-awareness will help you make more conscious choices in all these situations.

Self-understanding also helps you avoid something called *professional myopia.* Because we all have our own systems of meaning making, two people will not only see the same thing differently, but they will often see two different things altogether. If you are not aware that your point of view is only one of many, you may fail to consider other possibilities before making decisions about causes and solutions. For example, suppose you are an intern in a predominantly White high school and you hear a Black student complain of feeling out of place and uncomfortable, perhaps even unable to work to capacity. Part of the problem may be that certain features of the school itself, such as staffing patterns, menus, and recreational opportunities, are suited to middle-class White students. If you are not sensitive to these dynamics, you may describe the problem as one of adjustment or even cultural deprivation. Your efforts, then, will be aimed at helping the Black student feel better about a situation that may in fact be unfair or discriminatory.

When people are aware of differences between themselves and others, they sometimes confuse *difference* with *deviance.* They do not describe it as something different; they describe it as something wrong. You are not going to like every client, nor everyone you work with, but making the distinction between difference and deviance can help you accept and empathize with a wider variety of people. The tendency to assess difference as deviance can also lead to poor choices in using power and influence. If you do not see your perceptions as one of many possibilities, you may react to different perspectives by trying to change them. A trait assessed as a flaw in a client's personality, for example, may in fact be a cultural difference (Axelson, 1999). A family may choose to sacrifice some degree of their children's comfort and education to allow elderly grandparents to live in relative comfort. The family then becomes the target of a complaint by the school system. The family's decision may reflect a cultural value about the relative responsibility families have to children and elders. If you do not see their decision in this light, you might condemn the family for their treatment of the children, thereby damaging your relationship with them. Furthermore, you might then choose to concentrate on changing the family, never considering the option of urging the community to accommodate a wider variety of family customs and structures.

Finally, the more you understand about yourself, the better you will be able to recognize the stages of the internship as they happen. It will not be enough to know, or suspect, that you are in Stage 1 (Anticipation) or Stage 2 (Disillusionment). What does that mean for you? What strengths can you draw on and what personal traps should you avoid? To move through the stage you are in, are there aspects of yourself, as well as aspects of the internship, that you must confront and try to change? If you can discuss these issues in your journal and talk about them with peers, instructors, and supervisors, you will recognize the challenges sooner and move through them more smoothly.

It is important to realize that self-knowledge is not like other knowledge. Once you have learned a fact or a skill, you have it (unless you forget it or get rusty). But you are changing all the time, and your system of making meaning is changing with you. In addition, just as there is always something new to see and appreciate in a work of art, there is always more to know about yourself. So, self-understanding is a process to which you must become accustomed and committed. It takes work, but it will pay you big dividends.

Of course, self-understanding is an enormous area, one that requires a continuous commitment. We have chosen several topics within the realm of self-understanding to concentrate on in this and the next chapter. Our goal is to introduce you to these areas, to give you some things to think about, and to provide you with resources to explore these areas further. We are trying to summarize some complex theoretical information, some of which may be new to you. You and your instructor will need to decide which of these areas merit your time and energy. That decision may be made for your class as a whole, or different class members may choose to pursue different areas. In any case, we hope this chapter is a resource for you throughout your internship.

VALUES

A *value* is an idea or way of being that you believe in strongly, something you hold dear and that is visible in your actions. In the previous section of the chapter, we argued that self-understanding is important; that is a value for us. You may believe strongly in taking care of your family, in serving your country in some way, or in the existence of a deity. You are probably very aware of some of your values, especially ones that have been challenged, debated, or highlighted in the media (e.g., values about abortion, euthanasia, or freedom of speech).

Others, though, are so much a part of you and are shared by so many people in your life that they don't seem like values—they seem like truths. For example, both of us always placed a high value on punctuality and even assumed at one time that people who were habitually late were irresponsible. We have come to understand, however, that not all cultures share the same view of time. It is important that you clarify your values and give some thought to how you will respond when faced with people who do not share them.

Values permeate your life; we could never list all the important areas they touch. However, here are some that may be especially important in your internship. Answers to these questions, especially those that describe how you believe things "should" be, are important clues to your values.

Sexuality How do you feel about homosexuality? Bisexuality? Heterosexuality? Teenage sex? Premarital sex? Monogamy? Extramarital sex? Various sexual practices?

Family How should a family be structured? Are single-parent families OK? Should one parent stay home with the children? Should there be a "boss"? If so, who should it be? How should decisions be reached? Should grandparents or other relatives live with the family?

Religion How important is religion to you? How do you feel about religions other than your own? Would you ever do something that a leader of your religion said was prohibited?

Abortion How do you feel about abortion for yourself? For others? Are there circumstances under which you believe it is morally wrong? Morally justified? Should teens be able to choose on their own? Should the father of the child have a say?

Euthanasia Should a person be able to choose to end his or her own life? Under what circumstances?

Self-disclosure What kinds of things is it OK to tell someone you hardly know? For example, would you tell that person your financial problems? Your family difficulties? Your income? Which of these things would you tell a close friend? Are there things you would never tell anyone?

Honesty How do you define this word? Are there different kinds of lies? Is it ever OK to lie? If so, when and to whom?

Autonomy Do you think people should normally make their own choices and accept responsibility for their lives? Or do you tend to see people as responsible for one another? How large a role do you think fate plays in a person's life?

Work How hard do you think a person should work? What do you think about people who only work enough to get by, and no more? What about people who seem to have no desire to find a better job or make more money? Those who always push themselves to work harder, no matter what? How about people who don't want to work at all?

Acceptance How important is it to you that people from different cultural traditions be able to do things in a way that is consistent with that culture, even though it may be unusual by mainstream standards?

Hygiene How often do you think a person should bathe? Should a person use deodorant? What would you think of a person who showed up at your office with dirty hands? A dirty face? Dirty feet? Dirty clothes?

Freedom of Speech Do you think people's speech should be limited or restricted if it is hurtful to members of a cultural group, such as women or Muslims? What about statements disagreeing with U.S. military policy? With democracy as a form of government?

Time   How important is it for you to be on time? That others are punctual?

Alcohol   Is moderate use of alcohol OK? Is it ever OK to get drunk? How often? What about binge drinking—is it a dangerous practice? Harmless fun? Is it up to the individual? Do you think alcoholism is a disease?

Drugs   Are there some illegal drugs that you think are OK to use in moderation? If so, is it also OK to sell them? What are your thoughts about the use or abuse of over-the-counter drugs or prescription medications?

These questions may help you clarify some of your values. Think of how you feel about these subjects in your own life, and also how you feel about them in general. You might, for example, believe that you would not have an abortion, but that people should be able to choose that option for themselves. It is also important to ask yourself how strongly you feel about these values and whether you are open to changing them.

You might also want to think about how you came to have these values. Did you choose them consciously, after careful thought? Did they come to you from your family, your friends, or from a cultural group to which you belong? Did you accept them more or less uncritically? Perhaps you are not really sure where they came from and why you hold them. That is fine, but you may want to think critically about them before they are challenged.

You will almost certainly encounter someone at your internship who does not share some of your values, and you should think about how you might respond to that person. It might be a client, a coworker, or even your supervisor. Discussions about values can be lively and interesting if both parties are open to the discussion. But discussion is not always advisable or appropriate. If there are some values that you cannot or will not accept in a client, a coworker, or a supervisor, that is something you want to know as soon as possible and perhaps to discuss with your supervisor or instructor.

REACTION PATTERNS

Reaction patterns are ways that you respond—your thoughts, feelings, and actions—to particular kinds of situations. Some patterns are helpful and work well. Others are distinctly unhelpful; they do not appear to get you what you want, and yet you repeat them in spite of yourself. One of the most frustrating, stressful experiences people can have is one in which they find themselves doing something they don't want to do or reacting in a way they don't want to react. Here are some examples:

- A friend has seen your in-class presentation and offers some constructive criticism. As the conversation goes on, you find that you are getting angrier and angrier and are having a hard time listening. You keep coming up, mentally or verbally, with defenses for every criticism, and you imagine telling your friend off.

- You are at a party and people are talking about some hot topic. You have something to say but can't seem to say it. Since others are very vocal, it is easy, although frustrating, for you to just sit there. Later, someone says exactly what you were going to say, and everyone seems impressed.

- You are struggling with an intimate relationship. A good friend asks you how it's going with that person, and to your surprise, you hear yourself saying that things are fine.

- A friend calls late at night and asks that you to meet him right away. The matter does not seem like an emergency, but you leave your homework undone and go off to meet your friend.

These are not situations in which you later find, after much reflection, that you made a mistake. They are situations in which you know immediately afterward, or even during the situation, that you are not responding the way you want to. In fact, almost as soon as it is over, you can think of several ways to handle the situation that would have been better. Think about situations like this that have happened to you. Jot down a few of them.

You may find that these are isolated incidents or that they occur with just one person. You may also find, however, that these responses are part of a pattern for you. You may find that in general you are defensive about criticism, unable to speak in groups, unable to say "No" even when you want to, or to ask for help. Gerald Weinstein (1981) describes these tendencies as "dysfunctional patterns." Please note that, in this case, *dysfunctional* does not mean useless, nor does having a dysfunctional pattern make you a dysfunctional person. On the contrary, these patterns are clues to some important aspects of yourself. We both have them, and so does everyone else. The reason we are asking you to think about these patterns is that you may "bump into" them during your internship. We talk more about what patterns you might encounter and what to do about them in Chapters 10 and 11.

See if you can identify any such dysfunctional patterns in your life, and try using the format suggested by Weinstein:

Whenever I'm in a situation where _____, I usually experience feelings of _____. The things I tell myself are _____, and what I typi-cally do is _____. Afterward I feel _____. What I wish I could do instead is _____ .

Here is an example to help you:

Whenever I am in a situation where I feel angry at a friend, I usually experience feelings of anxiety and self-doubt. The things I tell myself are, "Take it easy. It's not that bad. There's probably a good explanation, and besides, you don't want to upset him." What I typically do is smile, joke, or protest very weakly. Afterward I feel as if I let us both down. What I'd like to do is find a clear, respectful way to tell my friend what is upsetting me.

And here are two from student journals:

I am trying to build my self-esteem and confidence. It is hard for me to hold back my emotions when something upsets me, especially given my fear of doing some-thing wrong or failing. I appear to be happy and cheery, but it takes so much out of me that I become negative and sometimes grumpy. This gets in the way of me liking myself.

I feel guilty that I need to rest and refuel my batteries, but this is my life. I dislike that I want to do everything perfectly.

LEARNING STYLES

One of the most important things you can discover about yourself is how you learn best. You have probably heard the labels "fast learner" and "slow learner" applied to other students or even to you. Recent research in educational psychology has shown these terms to be a bit clumsy. What seems to be true is that there are different styles of learning, and that some learning experiences require certain styles. If you are engaged in a learning experience that is not well suited to your style, it may be a struggle. Certainly, you can think of times when you learned something easily and other times when it was more difficult. Your internship is going to provide you with a variety of experiences to learn from, and some will be easier than others.

If you are involved in an internship that is not well matched to your style, you can still learn, but something needs to give. You may be able to take steps to change or augment the learning experience so that it is better suited to your strengths. On the other hand, you may need to try to stretch your learning repertoire and strengthen a style that does not come easily to you. In this way, a mismatch of styles can be an opportunity for growth.

Knowledge about learning styles can also help you be more effective with clients as you help them learn to do new things, such as change their behavior, locate resources, or improve their parenting skills. If you can learn to adapt interventions to a variety of learning needs, you will be more successful in this attempt. If you understand diversity in learning styles, you are less likely to view those different from your own as deviant or deficient.

You may have studied learning styles in some of your classes; there are a number of theories to draw upon. The variety of these theories can be exciting, but it can also be confusing. Claxton and Murrell (1987) have pointed out that many theorists use the term *learning style* to describe qualitatively different aspects of learning, thus making a single definition difficult. It would not be accurate, then, to speak of a student's learning style unless referenced to a specific theory, because learning style is not a single, static trait, but a combination of many characteristics and tendencies. We have chosen two areas in learning style to call to your attention. The first involves the way you take in and make sense of information; it draws on the work of David Kolb. The second concerns the way you relate to knowledge in general; it draws on the work of several psychologists and educators.

Kolb's Theory

David Kolb (1984, 1985) set forth a cycle of four phases that people should go through to benefit from experiential learning, as illustrated in Figure 3.1. The first phase is concrete experience (CE); students have a specific experience in the classroom, at home, in a field placement, or in some other context. They then reflect on that experience

FIGURE 3.1
Kolb's Experiential Learning Cycle

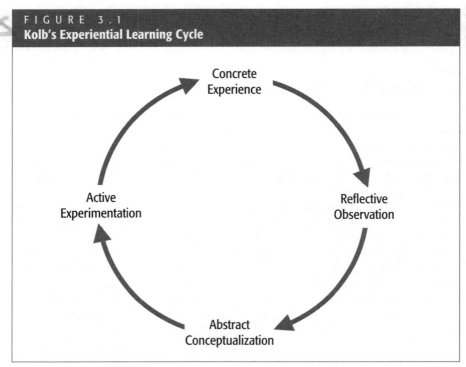

Source: Kolb and Fry (1975)

from a variety of perspectives (reflective observation, or RO). During the abstract conceptualization (AC) phase, they try to form generalizations or principles based on their experience and reflection. Finally, they test that theory or idea in a new situation (active experimentation, or AE), and the cycle begins again, since this is another concrete experience.

You may recognize this cycle from your internship. For example, suppose you observe a client in an argument with a staff member. You could draw on several theories or ideas you have learned to try to understand what happened, or you might seek out some new information from staff, books, or articles. You then begin to form your own ideas about what happened and why, and you might use this knowledge to guide your own interactions with that client. Once you do that, the interaction is itself a new concrete experience, and the cycle begins again.

Although the experiential learning cycle begins with concrete experience and continues in order, Kolb also points out that there are two fundamental dynamics involved in learning: taking in information and processing it in some way. Each of these dynamics can be approached in a concrete or abstract way. The concrete experience phase of Kolb's cycle is a way to take in information in a very concrete way—through experience. However, it is also possible to take in information in a more general, abstract way, such as through reading. This is abstract conceptualization. When information is processed in a concrete way, it is through experimentation (AE); when it is processed abstractly, it is through reflection (RO). You can get a good sense of where you fit in this model by taking Kolb's Learning Style Inventory, which you can order from the publisher listed in the References section, or you can find it online at www.hayresourcesdirect.com.

So now that you know where you fit in Kolb's theory, what do you do with that information? First, consider the match of your style to the demands of your internship. Most internships ask you to be active very soon after you begin, and to learn from your experience. This can be of great value, and some students who have struggled in more abstract classes feel much more competent and at home in the internship, where they are expected to learn by doing. However, if you tend to learn best in an abstract manner, you may feel a bit lost. When these mismatches occur, we encourage you to see what you can do to give yourself the kind of learning experiences you need. So, for example, if you are assigned a number of readings about various forms or procedures at your agency, that will be easier to learn if you prefer an abstract mode of taking in information. If you do not, you might learn the same information more easily by practicing on actual (concrete) cases.

However, you should also view the internship as a chance to stretch your learning style repertoire by becoming more adept at styles that you now find difficult. There is evidence that exposure to and competence in a range of learning styles help you become a more flexible and complex thinker (Kolb, 1984; Stewart, 1990), and Kolb's four phases can be thought of as essential abilities that all students need to develop (Sugarman, 1985). The concrete experience phase requires you to involve yourself fully, openly, and without bias in an experience, avoiding the rush to analyze and interpret. Active listening, which you may have studied, is a perfect example of these abilities. If a client or colleague is talking to you, it is difficult to really listen if you are busy thinking about what the hidden meaning might be or what you are going to say when the client is finished. Reflective observation requires the ability to consider an experience from a variety of perspectives. You must be able to slow down and apply several different perspectives (which may contradict one another) before forming conclusions. In your internship, you might be asked to analyze a case or a problem using a variety of theoretical frameworks. In abstract conceptualization, you begin to form theories or principles of your own. The ability to integrate is important here, as well as the ability to synthesize and generalize. In the active experimentation phase, you test your ideas, which means that you must be willing to take risks and make mistakes.

Knowing more about your learning styles and continuing to learn new theories on this topic will make you a more successful student, a better advocate for yourself and others, and a more effective intern. It will also help you set goals for self-improvement.

Separate and Connected Knowing

During the 1980s and since, a body of work in psychology has focused on the elaboration of two basic approaches to the world: separation and connection (Gilligan, 1982; Gilligan, Ward, & Taylor, 1988; Lyons, 1983). *Separation* is an approach to the world that emphasizes autonomy and abstract principles. *Connection*, on the other hand, emphasizes relationships and the importance of context. Much of this work was originally undertaken to fill a gap in developmental psychology caused by the use of predominantly male samples in psychological research and observation. By focusing on the experience of women, scholars were able to uncover important aspects of human development. Because of these origins, this work is often referred to as work on *women's development*. However, it appears that both orientations are found in men and women, although it also appears that the connection orientation is more often found in women.

The ideas of separation and connection were applied to learning in the book, *Women's Ways of Knowing* (Belenky, Clinchy, Goldberger, & Tarule, 1986). *Separate* and *connected knowing* are terms used to describe opposite ends of a spectrum. Most people show some evidence of both styles, and most also lean distinctly in one direction or the other. Each way of knowing has strengths and limitations. In describing these styles to you, we emphasize those aspects that seem most pertinent to your work as an intern. As you read the descriptions of these two styles, see whether you can recognize yourself.[1]

Separate knowers try to understand through analysis. They apply the principles learned in various disciplines. The scientific method, which you have probably studied, is an example of such a set of principles that is used to try to uncover scientific truths. You may also have learned methods for analyzing a poem, a play, or a painting.

Principles of analysis let us compare and evaluate. For example, two works of art may use different media, come from different times in history, and have been created in very different emotional contexts, but they can be compared and judged as works of art. Separate knowers, then, tend to separate themselves from too many specific details and look instead to organize experience according to abstract principles.

Turning now to the understanding of people, disciplines such as psychology and sociology offer separate knowers different lenses through which to make sense of human behavior. Each theory within these disciplines offers a different lens as well. Theories of human development (such as those of Piaget, Gilligan, and Erikson) as well as psychological theories (such as cognitive behaviorism, Gestalt, and transactional analysis) offer distinct ways of examining a person's behavior and reactions. A separate knower tries to understand a person by applying such theoretical frameworks.

Connected knowers, on the other hand, believe that truth is personal, specific, and located in experience. They attempt to understand not by analyzing but by empathizing. Rather than apply abstract principles, they immerse themselves in specifics. Connected knowers approach an idea or person with the belief that there is something of merit there, and they try to find it. In attempting to understand people, connected knowers try to share the experience of the person they are talking with or thinking about and understand what created that experience. They are less drawn to theories as a way of making sense of a person's experience.

Table 3.1 summarizes the key differences between these two ways of knowing and approaching the world.

---

[1]*Women's Ways of Knowing* is a complex and interesting book. It identifies five epistemological positions taken by women (and presumably by some men as well). The distinction between separate and connected knowing is made within one of those positions, called *procedural knowledge*, which is a position commonly found among college students. We urge you to explore this work further.

TABLE 3.1		
Two Ways of Knowing		
	Separate	**Connected**
Mode of Understanding	Analysis Abstract Principles	Emphathizing Immersion
Understanding People	Psychological & Sociological Theory	Experiential Logic of Each Individual
Relationship to Ideas	Skeptical Critical	Look for Merits

Perhaps you recognize some of your tendencies from these descriptions. Human services is an arena where connected knowing is valued, as opposed to being a liability, as it may be in many learning situations (Belenky et al., 1986). You will need to use both separate and connected approaches in your internship. What is important is to know where your natural strengths lie, what to use when, and where you will need to stretch and grow.

Many of your courses up to now probably emphasized the learning of principles and theories. In your internship, though, you are going to be immersed in experience and in specific situations. The theories you know will be only as good as your ability to use them in service to clients and the organization. We have known many students who sometimes struggle with theory but are outstanding in the field, and vice versa.

Theoretical knowledge is important in an internship. When you counsel a client, conduct an interview, plan a change strategy, or organize a project, theories help you know how to proceed. Connected knowers' tendency to want to "get into the head" of the theoretician may help them achieve a richer and fuller understanding of the theory. However, when it comes to using that theory to analyze a specific person or event, they may struggle.

In working with clients, a theory may help you understand, but it will not necessarily help you connect, nor to develop trust and empathy. After all, you are not interacting with a developmental stage or a psychiatric syndrome; you are interacting with a person. Most people wish to be seen as unique, not reduced to a stage or a diagnosis. The more you discover the experiential logic of a person (Belenky et al., 1986), the less strange that person's behavior and reactions will seem, and the better you will be able to communicate empathy and acceptance. On the other hand, if you use only the client's frame of reference, you lose the opportunity that theories provide to get a fresh look and to step away from repetitive, destructive patterns and rationalizations.

In your seminar class or support teams, there is a time and place for both separate and connected approaches. If you or someone else is struggling with a problem at your placement, a theoretical analysis can help you look at it in a new way. At the same time, a connected approach helps ensure that each person's unique experience is attended to and honored. Understanding the difference between separate and connected knowing may also help you understand your supervisor and coworkers better. Their expectations of and reactions to you may be based in part on their approach as separate or connected knowers. We will return to these themes later in the book.

FAMILY PATTERNS

For many people, their family of origin is the group with which they spend the most time until they form their own family or leave home for some other reason. Your experience in your family of origin is a powerful influence on who you are. Each family has its own way of doing things. Often, we are so accustomed to the way our family is that we assume that everyone's family is that way. For example, while both of us come from families where eating dinner together was very important, they differed considerably in how that time was spent. For one of us, conversation at the table was quiet and polite, and interruptions were frowned upon. For the other, the dinner table was lively, and often loud, with many conversations taking place at the same time. Imagine the shock for one of us visiting the other's family for dinner, or for either of us visiting a family whose members came and went from the table or ate in different parts of the house.

We are going to encourage you to think about two common features of family life: *rules* and *roles*. Although very few families have a list of rules posted on the wall, they all have unwritten rules that tell everyone what they can and cannot do. As a child, you can learn the rules by breaking them; someone reprimands or disciplines you, and you gradually figure out what is and is not acceptable. Sometimes families have rules that they are not even aware of. For example, one of us once listened to a family over the course of a weekend as they talked about all their relatives, living and deceased. There was one relative who was never mentioned, though. The family was quite surprised when this observation was shared, but upon reflection agreed that they almost never talk about that person. In observing this informal tradition, they were obeying a rule, even though they would never have called it that. Some family rules are a combination of individual family patterns and cultural norms (which we will explore more a little later on).

Here are some examples of family rules. Use this list to stimulate your thinking about the rules in your family.

- Don't talk about sex.
- Keep family business in the family.
- Never question a decision or disagree with an opinion from someone older than you.
- If your brother or sister picks on you, handle it yourself. Don't go crying to your parents.
- Guests are always welcome for dinner—you don't have to ask.
- Grandparents must always be consulted on important decisions.
- Mom and Dad need 15 minutes to relax after work before you ask them anything.
- No one is entitled to privacy, except in the bathroom.

Family roles tell you who in the family performs which functions. Some of them are pretty formal. One parent, for example, may pay the bills, do the cooking, or handle the discipline. There are other kinds of roles, though. There might be a family jokester, who is counted on to make people laugh, or the mediator, who tries to help settle conflict between family members. Some children get the role of the "good" child, and their mistakes are often overlooked or treated lightly, whereas others get the role of "bad" children and are treated more strictly. In some families these roles are shared, or certain people have them only in certain situations. Here are some questions to help you think about roles in your family:

- Who has the final say in an argument or dispute?
- Who is in charge when the parent or parents are not home?
- Which child is the smart one? The talented one? The athletic one?
- Who can you count on to help you out of a jam?
- Which child gets the most leeway from Mom? From Dad?

In your internship, you will undoubtedly meet people whose family patterns are different from your own. That can be pretty confusing as you try to make sense of their behavior and reactions. Understanding the sources of some of your behavior and feelings will, again, help you look more thoroughly and empathically at others.

Also, if you are not careful, you may carry into your internship a rule or a role from your family that is not helpful in that context. For example, your role as the jokester in your family, while it may serve to ease tension at home, may be annoying in a staff meeting, especially one where there is tension. Similarly, your role as the comforter may make it difficult for you to let a client struggle through his or her own problems.

PSYCHOSOCIAL IDENTITY ISSUES

You have probably had a course in developmental psychology in which you had some exposure to the ideas of Erik Erikson (Erikson, 1963). In fact, you may have had a lot of exposure to Erikson, and are now thinking to yourself, "Erikson? Again?" Well, yes, but just a little bit, and with a different twist. Although some of Erikson's ideas have been questioned and criticized since he first advanced them, some of the basic tenets of his work are valuable tools for self-understanding (see Sweitzer, 1993, for a full explanation). There are issues in each stage that are not always obvious from the familiar names of the stages (trust versus mistrust, and so on). Your early experience with these issues often reverberates into your later life, and we find that some of these issues are especially relevant in internships. Here, we will review those issues and encourage you to think about them and their relevance to your work. You may find that this theory does not seem relevant or helpful to you; there are lots of other developmental theories you can explore.

Trust vs. Mistrust: Getting Your Needs Met

Regardless of your level or preparation and skill, there will be times during your internship when you need help and support. Sometimes you will get it, and sometimes you won't. Your reaction to these events may be rooted in the Eriksonian stage of Trust versus Mistrust.

Children who lean toward a sense of basic trust believe that even though their needs are sometimes not met, fundamentally they can count on the world to satisfy their needs. Note that they do not believe that their every need will be met; children with that belief are in for disappointment. Children who lean toward a sense of basic mistrust, on the other hand, believe that even though their needs are sometimes met, fundamentally they cannot trust themselves or others to meet their needs.

For children with a sense of trust, a negative experience (one in which their needs are not met) does not change their fundamental view of themselves and the world. For children with a sense of mistrust, however, those events actually confirm the view that they cannot, and perhaps should not, have their needs met; they cannot trust their world.

You may be interested in exploring this part of yourself, trying to find out about your experiences as a child. What is more important, though, is exploring this part of you as an adult. What is it like for you now when you really need something and don't get it? No one likes that experience, but some people find it devastating. Or, you may be a person who doesn't ask for much from others because you are afraid you won't get it.

In your internship, you may find yourself needing things from your coworkers or supervisor. If they disappoint you, it is natural to be slightly upset, but a person with a shaky sense of trust may find these experiences unduly distressing. You may also find that, fearing your needs will not be met, you hide them, trying to appear confident and refusing to ask for help.

Autonomy vs. Shame and Doubt: Handling Your Impulses

Impulses and impulse control may seem like childhood issues, and indeed they are, although you will meet clients who struggle significantly in this area. But an internship is often powerful, emotional work in which you are not sure what will happen next, yet you have to react quickly. Sometimes you will handle these situations with ease and clarity. Other times you will find that your first impulse is to do something that in your calmer moments you would know is not appropriate. We all have these moments; how you react to them may be rooted in your experience in this stage.

In Eriksonian terms, children who end up toward the basic autonomy end of the continuum feel that even though they have to be controlled or curb their impulses sometimes, fundamentally their impulses are good, and they are capable of independent action. Instances where they are controlled (by others or by themselves) do not disconfirm their fundamental picture of themselves. Children who lean toward the other end of the continuum, basic shame and doubt, feel that even though they can sometimes do what they want—complete an impulse—their impulses are suspect and shameful, and they must not surrender to them.

Adults have impulses, too, and your experience in this Eriksonian stage may influence your feelings about impulses today. The issue is how you feel when you have an impulse, regardless of whether you act on it. Consider the example of an intern who is insulted in a particularly hurtful way by a client. Her first impulse is to insult him right back and put him in his place. She stops herself, though, and sets a limit with the client in a firm and calm manner. But her drive home, her journal, her conversation with peers, and her supervision time are dominated by how tempted she was. Rather than being proud that she controlled the impulse, she feels guilty for having it. The impulse, not the control, has confirmed her fundamental view of herself.

Even if this example does not sound like you, if this is a vulnerable area you may find yourself struggling with a particular kind of perfectionism at your internship. You may know how you should handle a situation, and you may even handle it that way, but you may have been tempted to behave otherwise. You may be tempted, for example, to be sarcastic, to call in sick when you are not, or to pretend you didn't see a client do something rather than have to confront it. Both of us have supervised interns who become upset with themselves not for what they did, but for what they almost did.

Initiative vs. Guilt: Finishing What You Start

Picture this: You are in a meeting at your internship and you offer an opinion on an approach to use with a group of clients. Your supervisor thinks it's a great idea and asks you, in front of the rest of the staff, if you would like to develop your idea further, design some activities, and co-facilitate them with another staff member. You say "Yes," but inside you are terrified, and can think of hundreds of reasons why this is a very bad idea.

Or, imagine you have an idea for a project in the community and mention it to your supervisor. Your supervisor thinks the idea has some merit, but does not think that the timing is right, or that you are quite ready to handle it. Later you find yourself angry and unable to let go of the anger. You feel like your supervisor is never going to "turn you loose" (even though this is really the first time you have asked). Both of these issues may be touching the Eriksonian issue of initiative.

Children with a basic sense of initiative believe that even though they sometimes may not be able or allowed to complete what they start, fundamentally they are capable of doing so and enjoy the inquisitive, adventuresome part of themselves. At the negative end of the continuum is a sense of guilt, which is felt by children who believe that even though they sometimes do complete what they start, their desire to explore and experiment is not a positive thing, and fundamentally they or the world will find a way to block these initiatives. The thwarting of an initiative, no matter how infrequent, confirms the fundamental belief these children have about themselves and reinforces feelings of guilt.

Ask yourself how you feel now when you have an idea, make a plan, or need to take charge of something. Does it feel like an adventure or more like a minefield? Think about your current reaction to being told "No" or having your desires frustrated in some way. Might some of these issues come up in your internship?

Industry vs. Inferiority: Feeling Competent

Your internship is your chance to shine, to see if you are really suited for work in this field or with this population. You may decide after the experience is over that it is not for you, but you want to feel like you can do it. A sense of professional competence is important to most human service professionals, and to most interns. Odd as it may seem, though, the trick is to be able to separate feeling competent from your successes and failures.

The internship offers countless opportunities for success and failure, and you will experience both. An interaction with a client, an attempt to defuse an argument, a group activity, a phone conversation, and more could all go awry. No one likes it when those things happen, but individuals with a sense of inferiority (in Eriksonian terms) may find them especially troublesome and difficult to overcome. You may also see clues to your position on this continuum in your reaction to criticism or evaluation.

Children who are praised only when they succeed, regardless of how much or little effort was required, may come to believe that success equals competence. Since no one can succeed all the time, their sense of competence may be tenuous; regardless of how much success they experience, they have a sense of basic inferiority. On the other hand, children who are praised for their efforts rather than the results, and who are encouraged to view failure as a learning experience and a natural occurrence, may develop a much more unshakable sense of competence. These children emerge with a sense of basic industry; they feel competent even in the face of failure. Children with a sense of basic inferiority, however, feel incompetent even in the face of success.

YOUR CULTURAL IDENTITY

One very important part of you, your feelings, thoughts, and behavior is your culture. The term *culture* is defined in many different ways, but for our purposes, we use the definition offered by Donna Gollnick and Philip Chinn (1990), that culture is a shared and commonly accepted set of beliefs, practices, and behaviors. They go on to point out that there are both microcultures and macrocultures. The *macroculture* consists of those beliefs and practices shared by the majority of citizens in that culture. So, for example, in the United States, a belief in democracy is shared by most (although not all) citizens and would be considered part of the macroculture. *Microcultures*, on the other hand, are those beliefs and practices that come from membership in a smaller group or subgroup. There are, for example, attitudes and beliefs that are more accepted by women than by men, and vice versa.

Your cultural identity consists of these subgroups, the degree of identification you make with them, your status as a member of dominant or subordinate groups, and your stage of subgroup identity development.

Subgroup Membership

You are a member of many subgroups. Some of them are temporary; you can move in and out of them as you choose. For example, you are now a college student, but you could stop being one at any time, and eventually you will no longer be a student. Other subgroups are relatively permanent; your membership in them is determined primarily by accident of birth. Your race, gender, and ethnicity are examples. Your age will change constantly, of course, but you will always be part of an age group. You are also part of an age cohort, sometimes called a generation. There are pieces of music and historical events that have enormous power for people of your particular age. Some have argued that sexual orientation is determined at birth; others believe it can change. However, your status as a heterosexual, homosexual, or bisexual is a relatively permanent feature of your life. Social class is another important subgroup, and here you need to consider both the social class in which you were raised and the one to which you now belong.

Degree of Identification

Knowing the subgroups you belong to is only part of the picture. Another important part is how strongly you identify with that group. There are Jews, for instance, who identify very strongly with their Jewish heritage. They observe the traditions and holidays, obey dietary laws, and travel to Israel. There are also Jews who do none of these things. Individuals make varying degrees of identification with the subgroups to which they belong. Although both of us are very aware of issues that confront us as a man or woman, we vary considerably in our identification with other subgroups. Fred, for example, knows almost nothing about his ethnic heritage, which is a mixture of Irish, English, and German, and is aware of very little influence from that subgroup. Mary, on the other hand, identifies strongly with both her Lithuanian and Irish heritage and has a good sense of how those traditions affect her as a person; however, she struggles to appreciate fully the effects of her race (Caucasian) in her personal and professional life.

Attitudes Toward Other Groups

Do you see yourself as a prejudiced person? Do you hold any stereotypes about people of a different gender, race, ethnicity, sexual orientation, or religion? If you are like most of our students, you will answer these questions "No" or "Not anymore." If you have qualified for an internship, you have probably been exposed to the core values of tolerance, pluralism, and respect for individuals that are at the heart of the helping professions. And naturally, you want to see yourself as a person who adheres to those values. If a part of you suspects that you do have some lingering prejudices, you may be keeping them hidden for fear that your peers or your professors will think badly of you or block your progress in your program.

Here is what we believe and invite you to consider. Everyone carries some stereotypes and prejudices, including us. If you work at it, you can learn to see your prejudices and make progress in overcoming them. However, at the same time, you will surely discover other, more subtle ones. The first step, though, toward being a non-prejudiced person is to confront and accept the prejudices you have. If you hide or deny them because you are ashamed or think that a good person would never have any, they will never change. Listen to what this intern had to say:

> Throughout my college years I held opinions of people, groups and cultures. I never once stopped to think about how my preconceived notions affected the way I acted towards people. I truly thought I was an open-minded person until I caught myself acting in this way.

Try this exercise. Pick a subgroup that has been the target of some discrimination and of which you are not a member. Depending on your own subgroup membership, some examples are Blacks, Muslims, lesbians, blue-collar workers, Native Americans, and many others. Now, think about all the stereotypes you know about that group.

Remember, we are not asking which ones you believe. Just write down as many as you can think of. You might want to get together with a classmate or two and expand your list. Now, look at your list. Where did the stereotypes come from? If you don't believe them, how did you come to "know" them?

Most students have little trouble coming up with a substantial list. The reason stereotypes are so easy to recall is that they are literally all around us. Think about the group you picked. How many members of this group do you actually know? In some cases the answer will be few or none. So where do your impressions of them come from? Some, of course, come from family and friends. Another part of the answer, though, is the way members of these groups are depicted in the media. Usually, members of various subgroups are portrayed in very limited, stereotyped roles. Think, too, about the books you read in school. How many members of these groups were in those books? How were they portrayed? The point here is that we have all "learned" harmful stereotypes about other groups (and even about our own).

Naomi Brill and Joanne Levine, in their excellent book, *Working with People* (2001), caution human service professionals to be sensitive to their use of "the paranoid 'they.'" If you find yourself thinking that "they" are too aggressive, too concerned with money, or too standoffish, that is a key to a stereotype or prejudice that you hold. We encourage you to explore and admit your prejudices and recommend some resources for doing so at the end of this chapter. Once you have identified some prejudices, you can work on changing them. For example, through an experiential exercise like the one we just encouraged you to do, Fred once discovered that he held some stereotypes about Hispanic men:

> I had to admit that I felt uncomfortable in groups of Hispanic men, and that I
> thought of them as being in general more violent and temperamental than me.
> Some reading about Hispanic cultural and family life helped me to get a more
> balanced picture of this group, with whom I had very little contact.

Mary recalls that she was somewhat apprehensive about working with same-sex couples in therapy. She attributed the discomfort to being generally uninformed about the homosexual culture:

> After reading a good deal and attending workshops, it became apparent that my
> discomfort stemmed from such stereotypical beliefs as gay individuals are more
> maladjusted, unhappy, and pathological than their heterosexual counterparts. It
> was only after an exploration of these beliefs that I could better understand the
> culture, as well as the basis for my own homophobic thinking.

As you get to know yourself, you need to consider your membership in various groups and subgroups in society and the positive and negative associations you have with your own and other subgroups.

Dominant vs. Subordinate

It may be that one or more of the groups you listed earlier is often referred to as a *minority group*. This term has been used not only to describe a group's relative numbers, but also its position in society. Because numbers do not always equal position (there are more women than men in society, for example), we prefer to use the terms *dominant* and *subordinate* to describe various subgroups (Hardiman & Jackson, 1997). Dominant groups are those who hold the majority of resources and power. They are more often in control of major institutions in society and are more often in positions of power in government, education, and business. So, for example, while women outnumber men in the general U.S. population, the vast majority of CEOs, members of Congress, college presidents, state governors, and so on are men. Thus, with regard to gender, men are the dominant group and women the subordinate. Using the same logic, in the United States, Whites are dominant with respect to race, heterosexuals with respect to sexual orientation, upper and upper middle class with respect to social class, and Christians with respect to religion.

If you think about it, you will see that it is a mistake to refer to a person as dominant or subordinate. At any given time, most of us are members of some subgroups that happen to be dominant, and others that happen to be subordinate. If you are a White, heterosexual, Jewish female from a working-class background, you are a member of two dominant groups and three subordinate groups. Go back to your own list of subgroups, and label each one as dominant or subordinate.

If you are hearing these terms in this context for the first time, you may have some trouble accepting them; many people do. After all, if you are a woman, you may not feel subordinate to men, and if you are White, you probably have no desire to dominate people of color. The terms do not necessarily describe your behavior. You are, however, a member of groups that, as groups, are relatively more or less powerful in society. More important for our purposes, in your internship you will be working with clients, coworkers, and supervisors who are members of different dominant and subordinate groups than you are. And in general, the experience of members of the dominant group tends to be pretty different from that of subordinate groups.

Dominant groups are afforded certain privileges by society. You may find that hard to believe, given all the publicity that affirmative action and other programs receive, but many of the privileges given to members of the dominant group are subtle. For example, both of us are White, heterosexual, and Christian. We can go into a department store and wander around as long as we want; friends of ours who are Black or Hispanic have told us that they are frequently followed by store detectives or questioned by clerks after only a few minutes. Wherever we have gone to school or to work, the major holiday in our religion is at least one day off for us and sometimes two weeks! When we are out in public with our partners, we can hold hands, hug, or even kiss without worrying about who may be watching or disapproving.

While members of both the dominant and subordinate groups may have stereotypes or prejudices about members of the other group, because the dominant group has more power, its members are more often in a position to enforce those prejudices, sometimes in very subtle ways. For example, one stereotype of women is that they are less assertive than men. For several years, Fred was an administrator at a residential treatment center for emotionally disturbed adolescents. Listen to what he says about his hiring practices:

> Although I would have told you I did not subscribe to that particular stereotype, looking back I see a difference in the way I looked at candidates for supervisory

positions. Discipline and assertiveness are very important in those positions, and while I hired both men and women, I realize now that I scrutinized the women much more closely than the men. It was as if the women had to prove to me that they could handle that aspect of the job, while the men had to prove they could not.

Because of these dynamics, members of subordinate groups are subjected to subtle, and not so subtle, acts of prejudice and discrimination. Some members of the dominant group perpetuate these acts deliberately (conscious oppression), while others are not even aware they are doing it (unconscious oppression). The incidents described earlier are more subtle. At the other end of the spectrum are verbal and physical harassment and physical assaults. The dynamics of dominant and subordinate groups are complex, interesting, and important. We encourage you to explore this topic some more. There are suggestions at the end of the chapter on how to do so.

Stages of Identity Development

If you are a member of a subordinate group, you may have gone through a period in your life when you didn't have much of a desire to interact with members of the dominant group. Or perhaps you never have felt this way, but have been puzzled by others who do. As a member of a dominant group, you may have been puzzled or hurt by this behavior. You may have thought of yourself as a non-prejudiced person and realized with some shock and shame that several of the things you and members of your group have done were harmful. All these attitudes and feelings are part of different ways that people make sense of their membership in dominant and subordinate groups. Interactions between members of various groups can be frustrating and confusing, and you may have experiences like this at your internship.

At the risk of introducing yet another theory, we want to tell you about some ideas that have been very helpful to us in sorting out these interactions. A number of authors have charted stages that people seem to go through as they grapple with these issues. Bailey Jackson, Rita Hardiman, and others at the University of Massachusetts have identified four stages that members of a subordinate group go through and four that members of a dominant group go through in the development of their subgroup identity (Hardiman & Jackson, 1997). We begin with the stages of the subordinate groups.

SUBORDINATE GROUP IDENTITY STAGES

Stage 1: Acceptance/Conformity   People at Stage 1 of subordinate group identity development either actively deny that there is any discrimination or passively accept it as "the way it is." They tend to believe that the dominant group occupies positions of power because its members have skills and abilities that they, as members of the subordinate group, don't have. They look to members of the dominant group for approval and validation and react strongly to any signals that they have done something wrong. Approval and support from members of their own group, while appreciated, are less important.

Stage 2: Resistance/Reaction   In this stage, members of the subordinate group often reject the norms and values set by the dominant group and try to build a positive sense of themselves as members of their own group.[2] Sometimes this involves an almost total rejection of the dominant group, accompanied by feelings of anger, pain, and distrust. A Latino, for example, may become more aware of discrimination at the hands of Whites, and dissociate himself with White society as much as possible. He may stop spending time with White friends, coworkers, or with Hispanics who do not see the world the same way, and may confront these people angrily over things they have done or said. He may also take steps to identify and spend time with the Hispanic community where he lives. Having grown tired and resentful of being punished or disapproved for not being White or not speaking, acting, dressing, and eating as a White person does, he may now seek to measure himself against standards drawn only from the Hispanic community and tradition.

Stage 3: Redirection/Redefinition   Here the process of building a positive identity continues, but now far less energy goes into rejecting the dominant group. Instead, the person concentrates on the movement, begun in the previous stage, to identify unique values and structures of the subgroup and to define him or herself as a member of that group. Although the person may withdraw from the dominant group, the angry rejection is gone. People at this stage tend to seek out others who are at similar stages in their identity development and draw support and validation from them.

Stage 4: Internalization   During this stage, the new identity becomes integrated. Because less energy is required to build and protect this new identity, people at this stage have more interaction with members of the dominant group, or with members of the subordinate group who are at different stages in their identity development. Membership in a particular subordinate group is seen as only one part, albeit an important one, of any person's total identity.

DOMINANT GROUP IDENTITY STAGES

Stage 1: Acceptance/Conformity   People at this stage of dominant group identity development can accept their status either actively or passively. Active acceptance involves conscious acceptance and approval of their superior status. People at this stage also discuss negative stereotypes of the subordinate group actively and appear to believe them to some degree. They deny that they are racist, sexist, or anti-Semitic (and so on), and insist that the subordinate group is to blame for its own problems in the world or is just not as capable in certain areas.

Passive acceptance is a bit more subtle. Here people are not aware of their discrimination and prejudice. They may concede that there is some racism or sexism, for example, but that it is blown out of proportion and that they themselves do nothing to support it. They are often offended and hurt by charges to the contrary.

Stage 2: Resistance/Reaction   Many people in the dominant group remain at Stage 1 for a long time or even throughout their lives. However, others begin to realize that there is prejudice and discrimination in society and in themselves. They see the negative effects that such discrimination has on members of the subordinate group, as well

---

[2]The term *subordinate* is used to distinguish between the power positions of subgroups. In this stage, while building a positive identity, members of the group probably do not refer to themselves as subordinate.

as on members of the dominant group, and often feel shame, guilt, or anger. This can be a very painful period.

Stage 3: Redirection/Redefinition   In this stage, the person rebuilds identity as a member of a dominant group, but in a different way. A man may define himself proudly as a man, for example, but not as a sexist. Actively working to eliminate sexism in himself and others, he now looks for the positive aspects of membership in his subgroup. He looks for other men at the same level to support him in his efforts and share his struggle.

Stage 4: Internalization   Finally, the new identity is fully integrated and, again, is only one part of the person's total identity. There is increased appreciation for diversity and for people at every stage of identity development.

It will take time to think about your own position on these stage continua. However, stage of identity is an important part of understanding why you react in certain ways to others and why they react to you as they do. We encourage you to read more about specific identity development theories. Recommended resources are at the end of this chapter.

Why have we devoted so much time to the issue of cultural identity? Many of the values, attitudes, and reaction patterns you examined earlier are strongly influenced by your membership in subgroups and the attitudes you learned about your own and other groups. Understanding the sources of these aspects of your personality will help you see them as only one of many possible ways to be in the world.

Furthermore, in your internship and your life, you will surely interact with people whose cultural identity is different from your own. You need to understand your reactions to them and theirs to you. Understanding cultural identity will help you accept others for who they are and avoid displacing feelings or prejudices you have about subgroups onto your clients and coworkers.

SUMMARY

This chapter has encouraged you to begin, continue, or develop the habit of self-examination, and we have suggested specific areas that will be important in your work. Making an investment in self-understanding will help you see your own issues as you move through the stages of an internship. It will also help you identify and overcome obstacles and deal more effectively with clients, supervisors, and coworkers. Perhaps you will also see the benefits in your life outside the placement. There are some other issues about yourself that you need to explore. These issues, which pertain primarily to your functioning as an intern, are the subject of the next chapter. For now, we leave you with a quote from John Schmidt (2002, p. 12):

> By seeking assistance for oneself and exploring our own development, we illustrate our belief in the helping process and demonstrate the same courage we expect of those who ask for our assistance.

For Further Reflection

FOR PERSONAL REFLECTION:
SELECT THOSE QUESTIONS MOST MEANINGFUL TO YOU

1. Review the areas of values discussed in the book. How many of them are areas about which you have strong feelings? Are there other core values for you that are not listed? Have your feelings about any of these issues changed over the years? Why do you think that happened? Discuss your answers with one or two other interns.

2. Which of these values do you think will be important in your internship? How will it feel to encounter others who do not share those values?

3. Think about how you prefer to learn. Do you prefer concrete methods of taking in information or more abstract methods? How about in processing information—do you tend to be concrete or abstract? Be sure to give specific examples. What learning activities and opportunities exist at your internship that are well suited to your style? Are there any that will stretch it?

4. Are you more attracted to and comfortable with using theories to understand people, as a separate knower does, or are you more likely to approach each person as an individual? Can you give examples of both separate and connected knowing in your life? How will your style be a strength and a liability in your internship?

5. Think about recent times when you have found yourself responding to a situation in a way you later regretted. Were any of these incidents typical of you? Do you think they are indicative of a response pattern? Are you aware of any other patterns that you struggle with? Try to write them out using the format on page 44. Which of these patterns might prove a challenge at your internship?

6. What are some of the important rules and roles in your family (both formal and informal)? How do they affect you? In what ways might they affect you in your internship?

7. Do getting your needs met, handling your impulses, finishing what you start, or feeling competent seem like particularly important areas for you? If so, do you think it may have something to do with the Eriksonian stages described in this chapter? How might these issues arise at your internship, and what will you have to be careful of?

8. Make a list of all the subgroups you belong to (race, gender, class, etc.). Next to each one, note whether it is a dominant or subordinate group and rate it 1–5 (low to high), depending on how strongly you identify with that group. Explain each of your ratings.

9. Are there subgroups in society about which you have some stereotypes? Will you be encountering any of these groups in your internship?

10. Choose one dominant group that you belong to. What stage of dominant identity development do you think you are in? Can you recall instances of being in previous stages?

11. Choose one subordinate group that you belong to. What stage of dominant identity development do you think you are in? Can you recall instances of being in previous stages?

SPRINGBOARDS FOR DISCUSSION

1. Take the Learning Style Inventory and share the results with your seminar peers. Discuss ways that you can complement one another's styles.

2. Share your thoughts about family rules and roles with your seminar peers. How might these rules and roles facilitate or inhibit a climate of mutual support?

For Further Exploration

ON THE IMPORTANCE OF SELF-UNDERSTANDING

Brill, N., & Levine, J. (2001). *Working with people: The helping process* (7th ed.). New York: Longman.

Excellent chapter on self-understanding.

Corey, M. S., & Corey, G. (2003). *Becoming a helper* (4th ed.). Belmont, CA: Brooks/Cole.

A very readable and thought-provoking book, devoted entirely to self-understanding and effectiveness in interpersonal relationships.

ON VALUES

Corey, G., Corey, M. S., & Callanan, P. (2003). *Issues and ethics in the helping professions* (6th ed.). Belmont, CA: Brooks/Cole.

The issue of values is woven throughout this book, but Chapter 3 in particular discusses the issue of imposing-versus-exposing values.

ON REACTION PATTERNS

Weinstein, G. (1981). Self science education. In J. Fried (Ed.), *New directions for student services: Education for student development* (pp. 73–78). San Francisco: Jossey-Bass.

Describes an approach to uncovering and interrupting these patterns.

ON LEARNING STYLE

Belenky, M. F., Clinchy, M., Goldberger, N. R., & Tarule, J. M. (1986). *Women's ways of knowing: The development of self, voice and mind.* New York: Basic Books.

An excellent and thorough treatment of this topic, with summaries of other research as well.

Claxton, C. S., & Murrell, P. H. (1987). *Learning styles: Implications for improving educational practices* (ASHE-ERIC Higher Education Report No. 4). Washington, DC: Association for the Study of Higher Education.

An overview of many different kinds of learning style theories.

Gilligan, C. (1982). *In a different voice.* Cambridge, MA: Harvard University Press.

This groundbreaking work laid the foundation for much of the work on women's development.

Kolb, D. A. (1984). *Experiential learning: Experience as the source of learning and development.* Upper Saddle River, NJ: Prentice Hall.

Kolb's basic ideas explained.

Kolb, D. A. (1999). *Learning style inventory.* Boston, MA: Hay/McBer Training Resources Group.

An easy-to-use guide to your own learning style and its implications.

ON FAMILY PATTERNS

Goldenberg, I., & Goldenberg, G. (2000). *Family therapy: An overview* (5th ed.). Pacific Grove, CA: Brooks/Cole.

Extremely readable text on families and family therapy.

Goldenberg, I., & Goldenberg, H. (1996). *My self in a family context: A personal journal* (4th ed.). Pacific Grove, CA: Brooks/Cole.

The accompanying workbook serves as a guide for you to explore your own family dynamics.

ON USING ERIKSON

Corey, G., & Corey, M. (2001). *I never knew I had a choice* (7th ed.). Belmont, CA: Brooks/Cole.

An excellent book on self-understanding based on Erikson and other theorists.

Sweitzer, H. F. (1993). Using psychosocial and cognitive behavioral theories to promote self-understanding: A beginning framework. *Journal of Counseling and Human Service Professions,* 7(1), 8–18.

Combines Erikson and reaction patterns and discusses implications for human service work.

ON CULTURAL IDENTITY

Adams, M., Bell, L. A., & Griffin, P. (Eds.). (1997). *Teaching for diversity and social justice.* New York: Routledge.

A collection of readings that describes workshops on many forms of oppression. Useful as a resource for trainers with exercises to work through.

Green, J. W. (1998). *Cultural awareness in the human services: A multi-ethnic approach* (3rd ed.). Upper Saddle River, NJ: Prentice Hall.

Covers both theoretical and practical aspects of working with diverse populations.

Ivey, A. E., Ivey, M. B., & Simek-Morgan, L. (1993). *Counseling and psychotherapy: A multicultural perspective* (3rd ed.). Needham, MA: Allyn & Bacon.

A relatively new version of a text that has been around for a long time. Unlike some other texts, where multicultural issues have just been tacked onto each chapter, Ivey et al. have fully integrated this perspective into a basic counseling text.

References

Axelson, J. A. (1999). *Counseling and development in a multicultural society* (3rd ed.). Pacific Grove, CA: Brooks/Cole.

Belenky, M. F., Clinchy, M., Goldberger, N. R., & Tarule, J. M. (1986). *Women's ways of knowing: The development of self, voice and mind.* New York: Basic Books.

Brammer, L. M. (1985). *The helping relationship: Process and skills* (3rd ed.). Upper Saddle River, NJ: Prentice Hall.

Brill, N., & Levine, J. (2001). *Working with people: The helping process* (7th ed.). New York: Longman.

Claxton, C. S., & Murrell, P. H. (1987). *Learning styles: Implications for improving educational practices* (ASHE-ERIC Higher Education Report No. 4). Washington, DC: Association for the Study of Higher Education.

Corey, G., Corey, M. S., & Callanan, P. (2003). *Issues and ethics in the helping professions* (6th ed.). Belmont, CA: Brooks/Cole.

Corey, M. S., & Corey, G. (2003). *Becoming a helper* (4th ed.). Belmont, CA: Brooks/Cole.

Erikson, E. H. (1963). *Childhood and society.* New York: Norton.

Gilligan, C. (1982). *In a different voice.* Cambridge, MA: Harvard University Press.

Gilligan, C., Ward, J. V., & Taylor, J. (1988). *Mapping the moral domain.* Cambridge, MA: Harvard University Press.

Gollnick, D. M., & Chinn, P. C. (1990). *Multicultural education in a pluralistic society* (3rd ed.). Columbus, OH: Merrill.

Hardiman, R., & Jackson, B.W. (1997). Conceptual foundations for social justice courses. In M. Adams, L. A. Bell, & P. Griffin (Eds.), *Teaching for diversity and social justice* (pp. 16–29). New York: Routledge.

Kegan, R. (1982). *The evolving self: Problem and process in human development.* Cambridge, MA: Harvard University Press.

Kegan, R. (1994). *In over our heads: The mental demands of modern life.* Cambridge, MA: Harvard University Press.

Kolb, D. A. (1984). *Experiential learning: Experience as the source of learning and development.* Upper Saddle River, NJ: Prentice Hall.

Kolb, D. A. (1985). *Learning style inventory.* Boston: McBer.

Kolb, D. A., & Fry, R. (1975). Toward an applied theory of experiential learning. In C. Cooper (Ed.), *Theories of group process.* New York: Wiley.

Lyons, N. (1983). Two perspectives on self, relationships and morality. *Harvard Educational Review, 53,* 125–145.

Schmidt, J. J. (2002). *Intentional helping: A philosophy for proficient caring relationships.* Upper Saddle Falls, NJ: Merrill Prentice Hall.

Schram, B., & Mandell, B. R. (2002). *An introduction to human services: Policy and practice* (5th ed.). New York: Macmillan.

Stewart, G. M. (1990). Learning styles as a filter for developing service learning interventions. *New Directions for Student Services, 50,* 31–42.

Sugarman, L. (1985). Kolb's model of experiential learning: Touchstone for trainers, students, counselors and clients. *Journal of Counseling and Development, 64*(5), 264–268.

Sweitzer, H. F. (1993). Using psychosocial and cognitive behavioral theories to promote self-understanding: A beginning framework. *Journal of Counseling and Human Service Professions*, 7(1), 8–18.

Sweitzer, H. F., & Jones, J. S. (1990). Self-understanding in human service education: Goals and methods. *Human Service Education*, 10(1), 39–52.

Weinstein, G. (1981). Self science education. In J. Fried (Ed.), *New directions for student services: Education for student development* (pp. 73–78). San Francisco: Jossey-Bass.

CHAPTER 4

Understanding Yourself as an Intern

> *If I can't help someone with a problem, I feel like a failure. I guess I feel like I should understand everyone and I should know what to do in every situation.*
>
> STUDENT REFLECTION

In the previous chapter, we encouraged you to examine who you are in general terms and discussed how some of what you discovered might be relevant to your internship. No doubt you discovered or rediscovered some issues that are relevant to your life and your placement and some that are not. In this chapter, we ask you to consider some specific issues that will be important as you begin your internship. Your motivation for your work, along with any unresolved or partially resolved issues that you bring to it, are important issues to consider. We also ask you to consider your feelings about self-disclosure, assessment, authority, flexibility, and reaction to dissonance, which we have found to be features of nearly every internship. Another issue that many interns don't think enough about is their life outside the placement, which surely affects and is affected by the work that they do. Finally, we ask you to examine and assess your personal support system, which will be an invaluable asset to you throughout your placement.

MOTIVATION FOR PLACEMENT

At some point in your education, someone has probably asked you to think or write about the reasons you are considering human services as a career. Perhaps, too, you have been asked by a placement coordinator or a prospective placement why you are interested in a particular kind of internship. Understanding and reminding yourself of these motivating factors can be a source of strength. Yet every one of them can be a liability as well. Corey and Corey (2003) discuss several reasons for entering human services, or needs that workers bring with them, including the need to have an impact, the need to care for others, the need to help others avoid or overcome problems the workers themselves have struggled with, the need to provide answers, and the need to be needed. They also point out that each one of these motivations can cause problems.

There is an important distinction between wanting and needing. All the reasons just listed are good ones for entering the human services field, provided you substitute *want* for *need*. For example, wanting to care for others is fine, but as Corey and Corey (2003) point out, it is not fine if you always place caring for others above caring for yourself. Sometimes what you need conflicts with what the placement wants or needs. For example, you have worked a very full week and are very tired. Your supervisor explains that someone is out sick and asks if you would work over the weekend. If you choose to say "Yes" to these requests from time to time, that is fine. However, if you find that you cannot say "No," ever, you may be someone who needs to put others' concerns first in order to feel good about yourself. That can happen for many reasons, but none of them offsets the price you will pay if you don't learn to attend to your own needs as well as those of your clients.

Providing answers is great, too; coming up with a solution for a problem is a wonderful feeling. However, there are going to be problems you cannot solve. Furthermore, there will be times when it is better to let someone struggle to find an answer than to jump in with a lot of advice, even if your advice is on target and would solve the problem faster. If you need to be the one with the answers, these situations are likely to be difficult for you. One intern realized in a journal entry, "There is some part of me that has a need to rescue these women."

If you have struggled with an issue in your life, such as depression, substance abuse, or an eating disorder, and feel that you have emerged from that struggle, pay special attention to your motivation as it relates to this issue. Often, past struggles are the reason people choose human service work (Collins, Fischer, & Cimmino, 1994). They want to help others deal with or avoid problems that they experienced and feel that their experience will be an asset. In many cases, it is. It can be a powerful motivator for you, and it can be a source of hope for both you and your clients. However, it can also be a trap. You may have learned a lot about your own struggle and your own path, but there is always more to learn about how to help others deal with the same issue. The techniques that worked for you may not work for others, and newer, more effective approaches may have been developed. You need to remain open to these possibilities. You also need to remember that your path is not necessarily the best path for each client. Just like people have different learning styles, they have different healing styles, and your job is to fit their needs. Your clients may need to struggle, just as you did, and it may be that you cannot and should not save them from that struggle.

Finally, as we alluded to in Chapter 3, some interns try to use the placement to resolve issues that they themselves have not yet resolved. If you encourage, or insist, that a client do what you cannot (e.g., be assertive or show love), or rush to protect a client from a critical parent (in a way that you never were), you are falling into this trap.

Take some time to think about your own motivations for entering the placement you have chosen. If those desires turn into needs, what kinds of problems could that create for you?

UNRESOLVED ISSUES

Each of us has struggled with different personal issues in our lives, and the self-examination you did in Chapter 3 may have uncovered or reminded you of some. There are other kinds of issues that can linger as well. You may have had a prolonged struggle with one of your parents during adolescence. You may have been a victim of abuse or assault. These struggles can leave us all with issues that are unresolved or partially resolved. For example, you may have overcome your adolescent struggle, but the memory of those days may still be very painful. These issues could be thought of as your "unfinished business."

This unfinished business does not have to come from some traumatic event or a struggle such as those just mentioned. Family patterns are another source. For example, if you were always the "mediator" in your family—the one who stepped in and calmed people down and helped them resolve their differences—you may have some very strong feelings about conflicts. It may be hard for you to see someone in a conflict and not step in to help, even though it is sometimes best to let people work it out for themselves.

As we mentioned in the previous chapter, your unfinished business can also come from your membership in certain societal subgroups. Your vulnerable areas are not just the result of your personality or of your childhood and family experiences. They are also the result of your experience in society as a member of racial, ethnic, gender, and other subgroups. If you are a member of a group that has been discriminated against, you may have a difficult time with clients who express prejudice, especially toward that group. If you are a member of a dominant group, such as males, Whites, or heterosexuals, and have thought about issues of discrimination, you may feel guilty, or very hesitant, around members of corresponding subordinate groups.

Even if your unfinished business is not part of the reason you selected human services as a career, human service work can often stimulate those issues. Suppose you are working with a substance abuse population. As you work with a client, he begins to discuss his struggle with an overcritical father. Even though you have not had experience with being a substance abuser, you have had a similar struggle with your father or some other parent figure. You are likely to be touched by this client in ways that seem mysterious at first. You may find yourself thinking about him when you are at home, or when you wake up in the morning. This kind of preoccupation can be very draining.

In addition, some clients have an uncanny sense for your sore spots and will use them to manipulate you. If you have a strong need to be liked, for example, the client may withhold approval as a way to get you to relax a rule or go along with a rationalization. Your unfinished business can show up with colleagues and supervisors as well. If your father or mother was hypercritical of you, you may decide that in the internship you are not going to accept criticism without standing up for yourself.

The point here is not that you should be free of unfinished business; all of us have some. However, you need to be as aware as you can of what that business is and how those areas of vulnerability may be touched in your work.

SELF-DISCLOSURE

All of us set boundaries around certain areas in our lives. There are aspects of ourselves we would discuss with anyone and aspects we would discuss only after we know someone for a while. An even more private "zone" consists of those things we would tell only our closest friends. There may even be things about you, such as events in your past or feelings you struggle with, that you would not discuss with anyone. How did these zones come to be private? Surely you have noticed that there are people who seem hesitant to discuss something that is perfectly comfortable for you or who are very open about aspects of themselves that you consider private. Perhaps your family taught you where to draw the line. Perhaps it was your culture or subculture. It may also be that you had experiences earlier in your life that left you cautious about trusting people with information about yourself. Regardless of where the boundaries come from, it is important to know where they are and how rigid they feel. At the end of this chapter is an exercise, adapted from the work of Paul Pedersen (1988), that can help you clarify these issues.

The work at your internship is likely to challenge your boundaries. Clients, coworkers, or supervisors may try to find out some personal information for a variety of reasons. Some may just want to get to know you better so they can feel comfortable with you. In some cultures, it is very important to exchange some personal information before addressing a problem or asking for help. Furthermore, going to someone for help, or especially being a client in a residential situation, can make a person feel powerless. Getting some information about you is a way to tip the balance of power a little more in the client's direction. Finally, some clients are actively looking to put you off balance.

You may feel hesitant to discuss a particular area but wonder whether you should try to overcome that reluctance. Or you may feel very comfortable but wonder whether certain topics are appropriate in a particular setting. For example, one study found that clients had little interest in the attitudes of their counselors but were interested in knowing about personal feelings, interpersonal relationships, professional issues, and successes and failures (Hendrick, 1990). In any case, you need to be prepared and think about what you will and will not discuss, and not just with clients. Self-disclosure can be an issue with coworkers, as you decide how much of yourself to share and with whom. Finally, self-disclosure is important in supervision. You will need to think about, and possibly negotiate, how much of your thoughts and feelings you want to share and what is expected of you.

ASSESSMENT

During your life, you have been assessed and given feedback in some form in many areas, including academic, artistic, and athletic. It probably began sometime during school. Most people get at least a little nervous when someone else judges their efforts, but for some people, it creates a great deal of anxiety. You may have been successful in some, or all, assessment areas in the past. You may also have a strong sense of competence, which you will recall is only partially related to success. If so, being assessed is probably relatively comfortable for you. However, it may be that you struggled in one or more of those areas or that you don't have a particularly strong sense of competence. Some of you have worked very hard to achieve the level of academic success you now enjoy. You have learned how to be successful in school and how to adapt to the different demands and styles of your professors. Even so, you are entering a new area now.

The attitudes, skills, and knowledge required for success at your placement site may be quite different from those required to be successful in classes. There is also likely to be an emphasis on what you can do rather than what you know. Put another way, you need to make use of your knowledge rather than simply show that you possess it (Kiser, 2000). If you have had other field experiences prior to your internship, the transition will be less dramatic, although each placement is different. For some interns, the internship feels like an area in which they have to prove themselves all over again (Kaslow & Rice, 1985; Lamb, Baker, Jennings, & Yarris, 1982). Think about your feelings and experiences with being assessed. Consider the prospect of a new round of assessment and how that makes you feel.

AUTHORITY

If you are working with clients, you may be in a position of some authority. For those of you who are parents, older siblings, former camp counselors, and so on, this is old news. For others, though, it will be a new experience, and it is one that causes many interns some concern (Blake & Peterman, 1985). Exercising authority isn't easy, and sometimes the people on the receiving end don't appreciate it. Being an authority figure can be especially difficult when your clients are close to you in age or have similar life experiences or circumstances. For example, we have had many interns of traditional college age who worked with adolescents just a few years younger than them. Their own adolescence was still fresh in their minds and with it the desire to test limits. Some had to struggle to feel comfortable enforcing agency policy with their clients. Take some time to think about your own experiences with authority, both as the authority figure and as the recipient of authority.

DISSONANCE

You may remember from Chapter 1 that an important step in Kiser's Integrative Processing Model is examining dissonance (Kiser, 2000). Part of the reason it is so important is that you will encounter lots of dissonance in your internship. It is important to stop and think a bit about how you typically respond to that experience.

You experience dissonance when two sources of information do not agree. You could encounter competing theories about group leadership, for example, or you might find that your internship uses an approach that is inconsistent with the theory you were taught and/or believe. Sometimes dissonance is resolved easily, sometimes it is resolved only with great effort, and sometimes it cannot be resolved at all. Your response to all three of these possibilities is shaped by your feelings about dissonance in general. Some people become so flustered and confused when theories disagree, or when theory and practice don't match up, that they ignore dissonance rather than have to deal with it. If they do see it, they may be more invested in getting it resolved quickly than in learning from it. On the other hand, some people are pretty comfortable with dissonance, and see it as a great opportunity for learning.

The ability to live with dissonance in cases where it cannot be resolved has also been called Tolerance for Ambiguity (Schram & Mandell, 2002). When things are not clear, but you have to act anyway; when you don't have any real way to tell how something will turn out; or when your clients leave your life without your knowing whether they have conquered the challenge that brought you to them, that is ambiguity. And human service work is loaded with it. However, some settings, and hence some internships, have more of it than others. Community change efforts often last far longer than the course of an internship. In client-focused settings, people who work with victims of domestic violence or who staff a suicide hotline are at one end of the continuum, while those who work in recreational settings such as the Boys and Girls Club or as employment counselors might be at the other. You need to think about your own tolerance level and the level of ambiguity that awaits you at your placement.

FLEXIBILITY

Another important aspect of your personality to consider is your level of flexibility. How do you react to sudden changes of plan? To multiple opinions? How important is it to you that your day or week is predictable? Of course, if you have chosen or are contemplating a career in public service, and especially in the helping professions, you are probably reasonably flexible. People are not easy to predict; that is what makes working with them challenging and fun, as well as draining or even exasperating. However, here again there are differences in internships, and you should think about how flexible you tend to be and how much flexibility is likely to be required of you. School-based helpers, while they have a lot of variety, at least have the structure of the school day to lean on. Crisis counselors and shelter workers probably have among the most unpredictable jobs.

YOUR LIFE CONTEXT

My biggest concern is stress, basically from outside the internship. The stress to get all the papers done on time, working many hours a week at my job to pay bills, and still put in 30 hours in the internship.
STUDENT REFLECTION

Your life context consists of all the other things going on in your life besides the internship. That context will vary according to your family situation, your social life, and the configuration of your academic program. You will have some responsibilities outside your placement, some expectations placed on you (or being continued), and possibly some stress. Here are some areas to consider.

School

In some programs, interns take very few classes during their internship semester. But in others, the internship can be the equivalent of only one course, leaving a full-time student with three or four other courses to carry. Some students even need to carry an overload because they must finish school within a certain time frame. The internship tends to be at least as demanding of your time as one course, or even two. Our students also report that internships demand a good deal of psychological and emotional energy, much more so than a regular course.

The work can be exhilarating, and many interns find themselves thinking about their work when they are not at the placement, or putting in extra time to read and research issues relevant to their work. But the work can also be emotionally draining, and the situations you face with clients and communities can be heartbreaking and frustrating. Many interns have told us that they find themselves thinking and talking about the internship a great deal, which sometimes makes it hard to focus on other classes.

Work

Some internships are paid; most are not. If you have been holding down a job in addition to going to school, you will want to think about whether you can and should give it up during the placement. Some interns cannot do this; their economic situation simply prohibits it. If you are going to be working, think about the schedule you will have and how it will fit into the time demands of your placement and coursework.

Roommates

If you live with other people, they are part of your life context. Schedules for housekeeping, sleep, quiet time, study, and entertainment are areas of negotiation with roommates, and you've probably already done that. However, you may need to do it again, depending on the demands of your placement. You may be up earlier, home later, or not available for cooking or other chores at your normal time.

Family

In this category we are including your family of origin (your parents, siblings, and anyone else who lives with them) and your nuclear family (your spouse or partner and children). You may be living with any, all, or none of these people, but you undoubtedly have responsibilities to them. What are they? How flexible are they? Could they be changed for this semester if the demands of the placement warrant it?

Intimate Partners

If you have a special partner in your life, whether or not you live with that person, this is another area of responsibility. It took time and energy to build that relationship, and it will take at least some time and energy to maintain and nurture it. How much time have you been devoting to that relationship? What does that person expect of you, especially now that you have begun your placement?

Friends

Although some friends are closer than others, every friendship is a responsibility. Think about the amount of time you spend with friends, individually or in groups, and about how much time you hope to spend with them this semester. Think, too, about the demands those friendships put on your time and energy.

Yourself

Don't forget to consider the time and energy you currently devote to taking care of yourself. By taking care of yourself, we mean sleep, good meals, exercise, and hobbies. All of us have activities we engage in for pleasure and as a source of stress reduction. Hiking, playing the piano, painting, lifting weights, meditating, and reading are just a few examples. You will need to take the best care of yourself that you can to meet the demands of the internship. You may be able to cut down some on hobbies if necessary, but you will need to be realistic about the time required to take proper physical and emotional care of yourself.

Doing It All

Many interns make the mistake of just inserting an internship into an already busy life and expecting it all to work out somehow. Some of our busiest students fall prey to this misconception, especially those who are raising families, working full-time, or seriously involved with sports. We encourage you to look a little more objectively at your life context. If you list all the people and activities that you will be committed to this semester and estimate the number of hours each week or month each of those commitments is going to take, you may be shocked to discover that you have committed yourself to more hours than exist in a week or a month. If so, just as when you are doing your financial budget, you will have to make some compromises. Often, this compromising involves talking with others to whom you now have commitments about reordering things, just for the short run.

Many interns report feeling that they have less time than they ever had. When we ask them to conduct an inventory of their commitments, as just described, they find that they should have enough hours in the day and week, but they just do not feel as if they do. Sometimes that comes from poor estimating. For example, if a class lasts for an hour, you need to remember that you will spend some time getting to and from class. One strategy you can try is to keep careful track of everywhere you go, how long it takes to get there, and how long you spend there for a day or two.

However, another reason for this feeling is that interns often underestimate the psychological commitment and emotional demands of the internship. Whether you realize it or not, your emotional energy is taxed by an internship, and that is going to affect how available you feel for other activities and responsibilities.

SUPPORT SYSTEMS

Balancing all the demands we just discussed is your support system, which is made up of people who give you what you need to get through life's challenges. Charles Seashore (1982), who has written extensively about support systems, calls them "a resource pool drawn on selectively in order to support me in moving in a direction of my choice that leaves me stronger" (p. 49). Your support system will be an important part of helping you meet the demands of your internship and the other demands in your life. As one of our interns pointed out in a journal entry, "Support teams are a way to relieve stress and frustration. . . . They listen to you, give advice, and give support. It is very comforting to know that you have someone to fall on when needed."

Everyone needs some support. Some people need more than others, and you will need more at some times than at other times, but you do need it and you will continue to need it. Sometimes, people in the helping professions have trouble accepting that they need support. Hence, they do not seek it out or they decline when it is offered. They give to others unstintingly and enjoy being people that others can count on, but they are better at giving than receiving. Most of the time, sooner or later, they become exhausted and are of little use to anyone. It may be that they have an image of the perfect helper—someone who never needs help. "How can I help others," students have asked us, "if I have problems myself?" Others know they need support, but they seem unable to accept it. For them, this may be a dysfunctional pattern of the sort described in Chapter 3. Accepting your need for support and developing a strong support system are important parts of becoming a human service professional.

Your support system is made up of many different people, and you will need different kinds of support at different times (Seashore, 1982). Here is a partial list of the kinds of support you might need. You may be able to add to it.

Listening Sometimes you just want someone who will listen to you without criticizing or offering advice. The person listening should be someone to whom you can say almost anything and on whom you can count not to grow restless or frightened. Think of these people as your "sounding boards."

Advice On the other hand, sometimes you need sound advice. You may not always follow it, but you need a source of advice that you can trust. Think of these people as your "personal consultants."

Praise There are times when what you need most is for someone to tell you how great you are. If they can be specific, all the better. Think of these people as your "fans."

Diversion Some people are friends you can count on to go out and play with. You don't have to talk about work, your problems, or anything else. You just have activities you enjoy together. Think of these people as your "playmates."

Comfort When we were children and we became ill, there was nothing we wanted more than pure comfort. A comfortable place to rest, good food, music we enjoy, all without having to lift a finger! At times, ill or not, this is still just the kind of support we need. Think of the people in your life who comfort you as your "chicken soup people."

Challenge   There are times when challenge is the last thing you want. At other times, though, someone who will push you to do more, look at things in a different way, and confront problems or inconsistencies in your thinking is the best friend you have. Think of these people as your "personal coaches."

Companionship   It is good to have people in your life with whom you feel so comfortable that you can do anything, or nothing, with them. Sometimes you may not care what you are doing, only that you are not doing it alone. Think of these people as your "buddies."

Affirmation   Another kind of support comes from people who have some of the same struggles that you do. It is a lonely and depressing feeling to believe, or suspect, that you are the only one who is troubled by a particular issue or set of circumstances. Knowing that others feel the way you do, even if they can't change it, can be very helpful. Sometimes there is no substitute for someone who has been through what you are going through. Think of these people as your "comrades."

As we said earlier, you are not going to need these things all at the same time. You are also not going to be able to get all of them from the same person or group. As you read the list, you must be able to think of people who are helpful with one thing but not with others. Both of us have friends who are fun to be with and provide wonderful diversion, but they do not listen well at all! Even if there were one person in your life who could provide all these kinds of support, that person would soon become exhausted if you asked him or her to meet all your support needs. Or, a person who usually can provide a particular kind of support may not be available at the time you need them. You will know soon enough if you have called on the wrong person for support. For example, you may need good advice, but find you have called on the listener. Try not to become frustrated. You need to learn about your needs and about how different people can best help you meet them.

Think about these categories and add some of your own. Now think about how well each of those needs is currently met in your life. You may find that there are some gaps that you need to address. Your support system will be stronger if you have more than one person in each category. Remember that the internship may, by itself or in combination with other things occurring in your life, create a greater need for support than you now have. It may also strain your existing support system. If most of your friends are not involved in internships, for example, they are not going to know how you feel, and they may not understand what you do or how difficult it is. "That's work?" they may say, or, "You get credit for that?" They may not understand your need to get to sleep early, work on weekends, or cut back on your social life. Or they may have negative attitudes about the population you are working with, like homeless or addicted people, or people with HIV/AIDS.

If you have discovered gaps in your support system, we urge you to take steps to fill them. Cultivate new friends; discuss your needs with friends and family. Like investing in self-understanding, it will pay off for you now, and later.

SUMMARY

We have asked quite a bit of you in Chapters 3 and 4. You have looked at yourself as a person and as an emerging professional through a number of lenses. Your assessment of your motivation, your unfinished business, and your response to several common internship issues took time and effort, but they are just as important to carry with you into the internship as any set of helping skills you have learned in class. We also hope you are now in the habit of considering yourself in the context of the people and systems that make up your life. Many human service students learn to assess clients in that way; we have encouraged you to shine that light on yourself as well.

For Further Reflection

FOR PERSONAL REFLECTION:
SELECT THOSE QUESTIONS MOST MEANINGFUL TO YOU

1. What are the major reasons you chose human services as a field? This internship in particular? Can you see any way that those motivations could be troublesome for you in your internship?

2. Have you struggled with a particular issue or problem in your life? How might that struggle be an asset to you in your current placement? What traps does it represent for you?

3. What unfinished business do you think you have at this time? How might that business arise during your internship?

4. In what ways and in what areas have you been assessed in your life? In how many of those areas were you successful? How do you feel about entering an arena in which you are relatively untested?

5. Take some time to think about your own experiences with authority, both as the authority figure and as the recipient of authority. How easy do you think this is going to be? Look back over what you have been learning about your life experience. Are there certain populations or issues that you may find difficult?

6. How well and under what circumstances do you respond to dissonance? See if you can come up with one example of handling it well and one where it threw you. How much dissonance are you anticipating in your internship, and in what forms?

7. How much predictability and structure do you think you will need from an internship in order to feel comfortable? Do you have any idea yet whether the setting you have chosen will meet those needs?

8. Review the part of this chapter titled "Doing It All." Make a list of all the people, activities, and other responsibilities in your life now (don't forget to include activities to take care of yourself, such as eating, sleeping, and exercise). Now make an estimate of how much time each week you need to devote to each one. Finally, add up the time. If it exceeds 168 hours, you are over budget—that's all the hours there are in a week. If you are over the limit, what do you plan to do about it?

FIGURE 4.1
Sources of Support

Category	Source 1	Source 2	Source 3	Source 4
Listening				
Advice				
Praise				
Diversion				
Comfort				
Challenge				
Companionship				
Other _____				
Other _____				

---

9. Using the chart in Figure 4.1, list the people you can count on in each category of support. There are spaces at the end to add your own categories. Now review your chart and see whether there are gaps—spaces with no one—or people who are in danger of being overburdened.

SPRINGBOARD FOR DISCUSSION

Here is an exercise about self-disclosure adapted from Paul Pedersen (1988). For each category listed, mark it as either an area you would talk easily about with relative strangers *(Public)* or one you would only discuss with those close to you *(Private)*:

	PUBLIC	PRIVATE
1. My personal religious views	_____	_____
2. My views on racial integration	_____	_____
3. My views on sex and morality	_____	_____
4. My tastes in food	_____	_____
5. My likes and dislikes in music	_____	_____
6. My favorite reading matter	_____	_____
7. The kinds of movies and TV shows I like best	_____	_____
8. The kinds of parties and gatherings I enjoy	_____	_____

	PUBLIC	PRIVATE
9. The shortcomings I have that I feel prevent me from achieving what I want	_____	_____
10. My professional goals and ambitions	_____	_____
11. How I really feel about the people I work and go to school with	_____	_____
12. How much money I make	_____	_____
13. My general financial situation	_____	_____
14. My family's finances	_____	_____
15. Aspects of my personality I dislike	_____	_____
16. Feelings I have trouble expressing	_____	_____
17. Feelings I have trouble controlling	_____	_____
18. Facts about my present sex life	_____	_____
19. My relationship with my intimate partner	_____	_____
20. My relationship with my family	_____	_____
21. Things I feel ashamed of	_____	_____
22. Things about me that make me proud	_____	_____
23. How I wish I looked	_____	_____
24. My feelings about parts of my body	_____	_____
25. My past illnesses and treatment	_____	_____

From A *Handbook for Developing Multicultural Awareness* by Paul Pedersen. Copyright © 1988 American Counseling Association. Adapted with permission.

We encourage you to share your answers with others. If you want to keep the answers anonymous, put a number or symbol on the paper that identifies it as yours and exchange papers with classmates. Discuss the similarities and differences you find. Where do you think those differences come from? How do you think your clients might answer these questions?

For Further Exploration

Corey, M. S., & Corey, G. (2003). *Becoming a helper* (4th ed.). Pacific Grove, CA: Brooks/Cole.

Good coverage on motivation for going into the helping professions and on unfinished business.

Seashore, C. (1982). Developing and using a personal support system. In L. Porter & B. Mohr (Eds.), *Reading book for human relations training* (pp. 49–51). Arlington, VA: National Training Laboratories.

Theory and application of support systems. A classic article.

References

Blake, B., & Peterman, P. J. (1985). *Social work field instruction: The undergraduate experience.* New York: University Press of America.

Collins, R., Fischer, J., & Cimmino, P. (1994). Human service student patterns: A study of the influence of selected psychodynamic factors on career choice. *Human Service Education, 14*(1), 15–24.

Corey, M. S., & Corey, G. (2003). *Becoming a helper* (4th ed.). Pacific Grove, CA: Brooks/Cole.

Hendrick, S. S. (1990). A client perspective on counselor disclosure (brief report). *Journal of Counseling and Development, 69,* 184–185.

Kaslow, N. J., & Rice, D. G. (1985). Developmental stresses of psychology internship training: What training staff can do to help. *Professional Psychology Research and Practice, 16*(2), 253–261.

Kiser, P. M. (2000). *Getting the most from your human service internship: Learning from experience.* Belmont, CA: Wadsworth.

Lamb, D. H., Baker, J. M., Jennings, M. L., & Yarris, E. (1982). Passages of an internship in professional psychology. *Professional Psychology, 13*(5), 661–669.

Pedersen, P. (1988). *A handbook for developing multicultural awareness.* Alexandria, VA: American Association for Counseling and Development.

Schram, B., & Mandell, B. R. (2002). *An introduction to human services: Policy and practice* (5th ed.). New York: Macmillan.

Seashore, C. (1982). Developing and using a personal support system. In L. Porter & B. Mohr (Eds.), *Reading book for human relations training* (pp. 49–51). Arlington, VA: National Training Laboratories.

DISCOVERING THE FIELD

Section One was the most theoretical and in some ways the most difficult section of the book. We have given you a lot of theory to understand and think about. Now we get to the business of applying that theory to your internship, moving through the stages of the internship, and helping you meet the challenges of each. The beginning of an internship can be overwhelming emotionally. Experiencing the internship is much different than reading about it or imagining it or even doing role-plays and simulations.

As you will see, the beginning of an internship can also be a very exciting time, with great intellectual as well as emotional challenges. There is much to think about and to prepare for. Section Two of the book will help you have as smooth a beginning as possible and set the foundation for further progress.

This section helps you focus on the tasks of the Anticipation stage. In Chapter 5, we focus on the major concerns that beginning interns often have about themselves, including acceptance, role clarification, and competence. We guide you in developing clear goals and objectives as you learn to write your learning contract. In Chapters 6 and 7, we help you look at your assumptions about and relationships with your clients, your supervisors, and your coworkers. Chapter 8 helps you become aware of and address concerns about your placement site as a whole from both a systems perspective and an organizational development perspective. And, in Chapter 9, you spend time thinking about the dynamics of the community in which your field work is being conducted and the relationships among that community, the field site, and the work you are doing.

CHAPTER *5*

Experiencing the "What Ifs": The Anticipation Stage

Many of us seem to have a lot of self-doubt about our capacities and what is manageable for us. It would seem that after all the required academics we would not have these feelings, but apparently not.
STUDENT REFLECTION

Y ou have probably been looking forward to your internship for a while. Some of you may have been waiting just a semester or two, others have been waiting their entire college careers, and some have been waiting a lot longer than that! You have heard about it, many of you have worked hard and sacrificed to get here, and now here it is. But as your friends and family ask whether you are excited, and you answer "Yes," you may also feel some anxiety creeping in. It may be a little or a lot; it may be visible to your friends, or you may keep it hidden. You may even hide your anxiety from yourself; some interns become aware of it only after it starts to go away (Wilson, 1981). In any case, you are entering an unknown experience, and that's always at least a little scary.

Stop a moment and think about some other new experiences you've had in your life. Do you remember your first day of school? Of high school? Of college? How about summer camp or a trip to see relatives you didn't know? An excursion into an unfamiliar neighborhood? You were probably excited about these things, too—and nervous. You have probably heard lots of stories from students who have already completed their internships; that is both a blessing and a curse. It builds your excitement, but you may also hear a little voice inside you saying, "Am I going to be as good at this as they are?" You may have heard some "horror stories" about internships that went sour in one way or another; even though these experiences are usually few and far between, the stories seem to have great staying power.

We call this first stage of an internship the "What if. . . ?" stage. Many interns express concerns during this stage. They wonder whether they will be able to handle it when _____ happens. Here are some of the "What ifs" we have heard:

WHAT IF

A client insults me?

A client confronts me?

A client gets physical?

A client attempts to approach me emotionally?

A client attempts to sexually assault me?

My writing skills aren't good enough?

I can't think of anything to say?

I don't learn quickly enough?

I get a disease?

I can't handle all my responsibilities at home and at the internship?

I don't have enough time for my loved ones?

I don't have time for those who need me?

A crisis happens and I don't know what to do?

A client lies to me?

A client falls apart?

I fall apart?

I make a mistake—and something really bad happens as a result?

Clients don't get any better?

Clients ask me personal questions?

I say something that offends a client?

I have a client I can't stand?

I lose my temper?

I don't grow up mentally and emotionally in time for graduation?

I have to discipline a client who won't listen?

I'm asked a question and don't know the answer?

The supervisor doesn't like me?

The other workers resent or ignore me?

My patient dies?

I get sued?

I hate it?

I don't feel my service is valid???????

Some of these may seem silly, and no one (so far) has worried about all of them. Your particular anxieties will be shaped by your personality, your knowledge of your placement site, and your past experiences. Every one of the entries on the list, however, is something we have actually heard from students. You can probably add to the list. We encourage our students to share their anxieties with one another; they are often surprised and relieved to find that they are not alone in their fears and concerns:

> *I have come to realize that we are all in the same boat although we are not all physically in the same sites. And, we are all going through the same feelings, worries and anxieties. Knowing this and seeing it in black and white has helped me tremendously.*
> STUDENT REFLECTION

Every one of those concerns is perfectly normal, and every one can be addressed, or at least the anxiety that sometimes accompanies them can be lessened.

THE TASKS AT HAND

Even though you may be a little nervous, you are probably eager to get going. After all, you didn't sign up for this experience to sit in a class, read a book, or write in a journal. You want to start doing the work you signed up for, acquiring new skills, and making a difference. However, at the same time as you start these processes, we are going to encourage you to attend to some other tasks:

- Examine and critique your assumptions.
- Prepare for the development of some key relationships.
- Acknowledge and explore your concerns.
- Set clear goals and objectives.

These tasks, if done well, will help address the concerns discussed earlier and will form a solid foundation from which to learn and grow. Without that foundation, you may falter sooner and harder.

Interns begin their placements with certain expectations, which are products of assumptions made, correctly or incorrectly, about many aspects of the upcoming or beginning internship (Nesbitt, 1993). As we mentioned in Chapter 2, these assumptions may come from stereotypical portrayals in the media of certain client groups or agencies, or they may be a result of your own experience. Your previous experience however, does not always predict what the future holds, although it is natural to generalize. You probably have at least some assumptions about clients, supervisors, the field site, and your coworkers. Making these assumptions explicit and subjecting them to critical examination will help you develop the most realistic picture possible of the internship you are beginning.

We said in Chapter 1 that relationships are the context or the medium for learning during the internship. You will be involved in many relationships with clients, supervisors, instructors, coworkers, and other interns during this time, and many of these relationships will be new. You are in the beginning stage of these relationships, and a normal concern at this stage is acceptance. No book can tell you how to achieve acceptance with all of these people, or even with any one of them. Each of these people is unique; furthermore, acceptance is an interactive process, and half of the interaction is you. All the facets of yourself that you explored in Chapters 3 and 4, and probably more, are part of the acceptance process. A relationship, then, is a process created by you and the other person, and as such is unique. However, there are some things we can help you consider and be aware of as you explore these new relationships.

One of the greatest concerns for beginning interns is their own competence (Brust, 1986; Yager & Beck, 1985). Naturally, you want to do a good job and be recognized for it. Furthermore, among all the professionals at work that you have observed, read about, or seen in videos, you probably have seen very few examples of mediocrity; the idea was to show you how effectiveness looks in action (Yager & Beck, 1985). However, a distortion effect can occur because of this. When we go to the movies, we see the scenes that were kept, not the ones that were scrapped or had to be reshot (again and again). Even the most brilliant baseball players fail to hit safely three out of five times!

Yet we never think about the failed attempts; we concentrate on the successes. Similarly, watching those videos and reading those cases make it easy to forget that everyone undergoes a learning process and that everyone makes mistakes. We imagine that we could never do quite as well, and we recall our own fumbling efforts. We often see only one of the paths to success and imagine we may never find it, while fearing the many paths to failure. So the question is, Can I really do this?

You may have been quite successful in your theory classes, and even in those that focused on skill building. Your previous field experiences may have gone well. But interns often tell us that the internship is "the big one," and some feelings are new to them. Some interns, for example, struggle with what has been called the *imposter syndrome* (Clance, 1985). They are vulnerable to believing that whatever success they have achieved is due to good fortune rather than competence. Every step of the way, they fear being exposed for the pretenders they imagine themselves to be. In the case of the internship, they imagine that they did well enough to qualify only through remarkably good luck and that they will surely be found out at their placement site. These are natural concerns; the feelings about assessment, competence, and initiative that you explored in Chapters 3 and 4 will shape how these concerns manifest for you.

The good news is that there is a cure. The bad news is that, in large measure, the cure is time. You will almost certainly feel more confident and competent as time goes on. In the meantime, if you check with your peers, you will see that you are not alone; that realization can have remarkable healing powers.

Another concern for many interns is the nature of their role in the field experience. We have explored this issue further with students; it seems that the issue has two parts. First of all, interns are concerned about how they will be seen by others at the site and how they will come to see themselves. Are you a student? A volunteer? An observer? A researcher? A staff member? In fact, you will probably perform all of these functions, but none of these roles by itself defines who you are at the internship. The role of an intern is unique. Initially, you may spend a good deal of time observing. Later you will be a more full-fledged participant, but it is important to retain some of your observer status (Gordon, McBride, & Hage, 2001). Because you are there to learn, not just to work, you must reserve some time to engage in and complete the experiential learning cycle we discussed in Chapter 3. You must take time to observe and reflect as well as take action and have experiences.

There may be volunteers at your internship site, and that can cause some confusion for you as well as for the staff. Volunteers and interns do some of the same tasks, but their roles and responsibilities are very different. Volunteers are essentially there to help out. You are there to learn. You have specific learning objectives the volunteers do not. As a learner, you should be doing, but you also have the responsibility to reflect, analyze, and critique, which a volunteer does not have. Furthermore, although you may perform some of the same tasks as a volunteer, as an intern you have a responsibility to learn why a particular task is done, why it is done in a particular way, and how it relates to the bigger picture in the organization (Royse, Dhooper, & Rompf, 2003).

Second, interns wonder what they will be asked to do. Will they have to dive right in? Will they be given boring, mechanical tasks like copying, filing, answering the phone, or driving? Your learning contract, negotiated with your instructor and supervisor, should help allay some of these fears. However, you should expect to be given some of what is often called *grunt work.* In many agencies, almost everyone pitches in and does some of the drudgery (Royse et al., 2003). You will probably be asked to do your share. You may even be asked to do a little more than your share so that the more skilled staff can focus on other kinds of work. It should be clear, though, to everyone concerned exactly who can assign tasks to you. If you feel you are being asked to do too much of this sort of work or are being asked by too many different people, you should speak with your instructor and/or your supervisor.

However, as Bruce "Woody" Caine (1994, 1998) points out, there is quite a bit you can learn from performing these tasks. Often, the learning is not in the task itself, but in the organizational meaning of the task. You get a chance to learn what it *really* means to be an entry-level worker in that agency. You become familiar with these tasks, which are critical if not always interesting, and how long it takes to complete them (which is easy for people who don't have to do them to forget!). You can also learn how much it costs the organization, in material and human resources, to get this work done.

THE LEARNING CONTRACT

I never thought that I would feel this way, but I am very happy I took the time to work on the contract . . . [because] it helped enormously to define my role.
STUDENT REFLECTION

At some point in the internship or the placement process, you, your site supervisor, and your field instructor will negotiate the goals of learning for your field experience. This written agreement is a document that articulates in detail what you will be learning, how you will go about learning it, how you will determine your progress, and how you will be evaluated on achieving your goals. This document may be referred to as a learning plan or agreement, depending on your academic discipline or campus policies. We refer to it as the *learning contract*.[1]

Examining the Contract

The contract perhaps is best described as a master plan for your field placement. It is your map to success in the field. Developing the right kind of contract is essential to maximizing your learning and ensuring that your learning needs are going to be met. For one thing, the learning contract minimizes the chance of misunderstandings. It also will keep you focused on your internship commitments and provide you with a sense of achievement. Most important, the contract will ensure the integrity of your academic work (Royse, Dhooper, & Rompf, 2003).

Some of you may be feeling a bit lost in your internship right now, not really knowing what you are going to be doing over the course of the field placement, or perhaps feeling the work you will be doing is insignificant to you or the site. If that is the case, it may be a relief to know that the learning contract will clarify your role and responsibilities, and you will have the opportunity to make sure that your work is meaningful to you and to the site. That translates into your having a sense of purpose in your field work. Without a sense of purpose, you will begin to question your value to the site. If you question your value as an intern, your overall morale and investment in your internship may be affected, and not in positive ways. We cannot stress enough how important it is for you to be involved in developing your learning contract and to work conscientiously on it, as doing so ensures that you'll have a most challenging and rewarding internship. In the next few parts of this chapter, we talk about the basics you need to know as you work on the contract.

The Writers of the Contract

There are three key individuals involved in creating this master plan: you, your on-site supervisor, and your campus instructor. Most experts in the field agree that an internship is most effective when all three parties work together to develop a field experience that meets everyone's needs and provides an opportunity for all to learn (Hodgson, 1999). What that means is that it is important that you are actively involved in determining your learning goals. Doing so may be intimidating at first, as this is probably a change in how you have experienced your academic courses up till now (Kiser, 2000). However, our experience tells us that once you become involved in the process, you will get beyond these feelings and share in developing your learning goals.

---

[1]There are two contracts that are usually involved in an internship. The first is an *internship agreement*, which is a legal document between the campus and the placement site; this contract details the responsibilities of the involved parties and describes in general terms what people will do to carry out the agreement. The second is the *learning contract*, which is a way to make clear to all parties what you want and expect to learn, and in what ways that will happen; this contract includes your specific learning goals and objectives.

The Essence of the Contract

So, what is it about the contract that makes it so important to the overall success of a field placement? There are three general components to the learning contract, and they are pretty straightforward: (a) learning goals, (b) activities to reach the goals, and (c) assessment measures. Some learning contracts also include behavioral objectives, which we will discuss in more detail later in the chapter. We have found that how well these components are developed and matched to your expectations and learning styles makes all the difference in whether you have an empowering and transforming field experience or a mediocre one.

In addition to these components, there are a number of details that need to be included in the contract. For example, you will want to be sure to note these points. (Rogers, Collins, Barlow, & Grinnell, 2000; Royse, Dhooper, & Rompf, 2003)

- The time spent on a weekly basis in the internship (days per week, hours per day)
- Supervision specifics (methods; individual/peer/group; frequency; length of time)
- Case specifics (numbers, types, time frame)
- Documentation and recording procedures
- Special arrangements
- The date the contract was drafted.

The contract itself should be demanding and ambitious, stretching you while enhancing your knowledge. It should be grounded in realistic time frames as well as in realistic expectations of the work and your capabilities (Horejsi & Garthwait, 2002; Kiser, 2000). Keep in mind when negotiating your contract that the goals you set for the end of the placement should seem lofty at the time you're writing them, and feeling intimidated by them is not uncommon. Be reassured, though, that the feelings of intimidation will pass and you will attain your goals, but it will take time and work to make that happen.

When developing your contract, remember that it is best if the contract is written so it can adapt to changes in the internship. In essence, the contract should be a living document (Kiser, 2000), not one carved in stone. Just imagine what would happen if an opportunity for an exciting project developed after you negotiated the contract and your contract was non-negotiable. You could miss out on what might be *the* defining aspect of your internship! It is important that you are able to renegotiate the contract to ensure the best fit with your overall learning goals. By the same token, if a project you agreed to take on does not materialize, you should be able to replace it with a more viable learning opportunity.

There are three factors in a contract that we consider in order to maximize the likelihood of you having a fulfilling field experience (King, 1988). First, your work in the field must be *worthwhile* for you to do it. That translates into you being given opportunities to accomplish tasks, be acknowledged and respected for your work, create your own work, push your limits to touch your potential, and enjoy what you are doing. Second, your supervisor and coworkers must be capable of demonstrating *responsible relationships* in working with you. That translates into competent supervision and collegiality. And, last, your supervisor must be willing to cultivate your *active and conscious involvement* in the internship. That translates into encouraging you to take on responsibilities and have an assertive role in defining your field experience.

The Timing of the Contract

The time at which you, your instructor, and your supervisor negotiate the contract tends to be a matter of policy—either of the academic program, of the campus, or of the field site. Sometimes, it becomes a matter of debate. There are programs that develop their agreements before the field placement begins, or sometime between when the internship agreement is negotiated (the site/campus contract) and when the internship begins. In programs with year-long internships, students can have up to a month in which to develop their learning plans, while interns in programs of a semester or less typically have from 1 to 2 weeks to develop the contract (Kiser, 2000).

Many experts agree that it is important that you have some understanding of your field site, your clients, and your learning tasks before negotiating the contract (Rothman, 2000). We develop our contracts within the first 3 weeks of the placement, regardless of the length of the field experience or hours in placement. We believe you can take an active and informed role in the contract process by the third week in the field; by doing so, you enhance your responsibility in the learning process.

The Levels of Learning

Part of the process of developing a learning contract is paying attention to the different levels of learning that need to take place along the way (King, Spencer, & Tower, 1996). Learning takes time, and your activities must be organized so as to build on your attitudes, skills, and knowledge as they develop and change over time. We find it useful to think in terms of three sequential levels of learning: (a) orientation, (b) apprenticeship, and (c) mastery.

The *orientation* level is a time when you become acquainted with the work of the internship. Days at the site are marked by lots of observing, reading, shadowing, and inquiring. These are far from aimless activities. You need to have goals during this time because you are building the foundation for a successful internship (Feldman & Weitz, 1990). The *apprenticeship* level is an intensive teaching and learning period, with much learning-by-doing under very close supervision. The third level is a period of *mastery*, during which you will be engaged in your own work and actually structuring and organizing your schedule to meet time lines and agency goals. You are more self-directed at this level of learning, and your need to understand subtle aspects of the work and yourself in relation to the work is heightened. Typically, you will need less direction and may or may not need much support from your supervisor.

Developing the Contract

There are many kinds of learning contracts, ranging from simple goal statements that apply to all internships to highly specific and individualized plans (Wilson, 1981). Your academic program may already have a format for developing learning contracts. If so, it probably takes the form of one of two general approaches: a *goals-focused contract* or an *objectives-focused contract*. Both of these approaches begin the contract process by identifying the *goals of learning* and end the process by setting forth *assessment criteria* to determine how well the goals are met. Both approaches also identify *activities* as ways of achieving the goals. Differences between the approaches arise in whether or not *learning objectives* are used to guide the learning contract (Horejsi & Garthwait, 2002; Kiser, 2000; Rothman, 2000).

It is generally accepted that the more succinct approach—the goals-focused contract—is easier to implement and provides sufficient direction to be a worthy document for students and agency personnel to use (Horejsi & Garthwait, 2002). In the paragraphs that follow, we discuss these components and give examples of each to guide your thinking.

DETERMINING THE GOALS

Goals are good. They work. As a matter of fact, there is substantial research that indicates that if you develop specific and challenging goals, your performance will improve (Curtis, 2000). A clear sense of goals also will help you focus your efforts. We know that if you actually write down your goals, you are more committed to achieving them (Caine, 1998). The goals of the learning contract are academic outcomes, not activities, and tend to fall into four general categories in the helping professions: (a) *knowledge*, (b) *skills*, (c) *development* (personal, professional, and civic), and (d) *self-assessment*. The following references were used to develop this section of the chapter: Gordon, McBride, & Hage, 2001; Horejsi & Garthwait, 2002; Kiser, 2000; Rothman, 2000; Royse, Dhooper, & Rompf, 2003; and Sgroi & Ryniker, 2002.

Knowledge Goals refer to learning and understanding factual information, terminology, principles, concepts, theories, and ideas of the profession. Typically, they involve learning new information, such as:

- Become familiar with how the criminal justice system in this state operates
- Learn the network of services in this city available to people who are physically challenged

Skill Goals refer to effectively demonstrating the behaviors of the profession and typically describe things you want to learn to do, such as:

- Learn to write a case report
- Develop oral presentation skills for courtroom work

Personal Development Goals refer to learning more about yourself as an emerging professional and identifying ways in which you want to grow and change. Although most interns undergo a good deal of personal growth during the field experience, they never anticipated much of what they describe. Here are some examples:

- Become familiar with a style for speaking in front of large groups
- Develop more confidence in working with the children of elderly clients

Professional Development Goals refer to learning the values, attitudes, and ways of the profession, such as ethical and legal awareness, deportment, record keeping, and disclosure. Typically, these goals focus on ways you want to enhance your professional skills and plans. For example:

- Develop intake interviewing skills
- Develop one-to-one counseling skills with adolescents

Civic Development Goals refer to learning the values you ascribe to citizenship and ways in which you can take more responsibility to be actively involved. These goals typically focus on learning more about yourself in relation to your local, state, and national governments. For example:

- Become familiar with opportunities for grassroots involvement in citizen advisory groups to state social service agencies
- Become familiar with community group involvement in restorative justice programs

Self-Assessment Goals focus on ways to measure your progress and goal attainment objectively. For example:

- Learn about a variety of self-assessment techniques, such as videotaping or otherwise recording my work as it is happening
- Learn to accept feedback from supervisors and coworkers

When writing the goals into the contract, include several for each category of learning (Kiser, 2000). The goals should be as specific as possible, especially if you are not using the objectives-focused contract, because the learning goals provide the detailed description typically found in learning objectives.

Regardless of the approach you use, learning goals are critical to the success of your contract and field experience. Developing goals can be tricky business. One technique that we find useful is sentence completion exercises like these.

For each category (knowledge; skills; personal, professional, and civic development; and self-assessment) answer the following questions:

- If I could make it happen, what goal would I like to achieve by the time I complete my internship? (deShazer, 1980, cited in Curtis, 2000)
- If I were given a magic wand, what would I like to see happen in my internship in the area of _____ goals?

Essential to formulating a goals statement is knowing the language to use to frame the learning goals. The following verbs are frequently used in goals statements and may help you in writing your contract (Horejsi & Garthwait, 2002, p. 21).

acquire	comprehend	learn
analyze	develop	perceive
appreciate	discover	synthesize
become	explore	understand
become familiar with	know	value

STATING THE OBJECTIVES

The decision to use learning objectives in the contract is often discipline specific. For example, social work programs tend to include objectives (Rothman, 2000); human service programs tend not to use objectives in their contracts (Kiser, 2000). Regardless, it is important to know how to use learning objectives appropriately in order to understand why your contract could look substantially different from a peer's contract even if you both are doing the same work on-site.

Learning objective statements are behavioral in nature and tend to be more detailed, concrete, and precisely worded than goal statements. Learning objectives describe specific actions and activities, for example, to demonstrate the ability to match needs of clients to available resources.

It has been our experience that developing learning objectives is both an art and a skill. The art is reflected in the vision of the objective; the skill is reflected in the ability to incorporate the goals, activities, and assessment measures in one statement and in realistic and heuristic ways. The following verbs typically are used in writing learning objectives (Horejsi & Garthwait, 2002, p. 22). If you are using the learning objectives contract, this list may be helpful.

answer	decide	obtain
arrange	define	participate in
circulate	demonstrate	revise
classify	direct	schedule
collect	discuss	select
compare	explain	summarize
compile	give examples	supervise
conduct	list	verify
count	locate	write

In stating a learning objective, the manner in which you plan to assess reaching the objective must also be included, along with a time frame for achieving it. Thus, the learning objective brings together (a) a goal, which is stated in more broad and general terms, (b) activities to attain the goal, and (c) assessment measures to determine if the goal was reached—all in a single statement. Not an easy task! A word to the wise if this is the approach you are taking: The integrity of your learning is critically important to the quality of your internship. If you choose to focus on learning outcomes that are easily measured rather than on important learning outcomes, you will compromise the quality of your field placement as well as the integrity of your academic experience (Horejsi & Garthwait, 2002, p. 22). Working with your on-site supervisor and with your field instructor in developing the contract will ensure that your learning contract is appropriate to your needs.

Rothman (2000) offers excellent examples of how to write and not to write learning objectives, advice that will serve you well. For example, she frames the contract in such a way that you move smoothly from the general (overall goals) to the specific (objectives) to the most specific (approaches, tasks, and methods), and then she suggests that you reverse the process when it is time to review your progress (p. 95). Rothman also stresses the importance of using measurement criteria that are both relevant and measurable (p. 99), and she recommends that your objectives be grouped by time frame so that you can easily review what you have done in the designated time periods. She offers an excellent example of a statement using both relevant and measurable criteria: the "ability to use empathy appropriately as demonstrated in five instances recorded in process recordings of three client interviews in one month" (p. 99). This example will guide you well as you are developing assessment objectives.

MAKING YOUR CHOICES

Strategies, activities, methods, and *approaches* are all terms to describe ways of accomplishing the goals or objectives. Sometimes they are used interchangeably. Technically, objectives are *strategies* for achieving goals. The ways should be noted as specifically as possible, describing what you will do, where, when, and how. As you probably suspect, there is much room for personal preferences when selecting ways to meet your goals; just be sure the resources are available.

An important aspect to consider when you make your choices is your *learning style.* Spend some time thinking about what style of learning works for you; then be sure to incorporate activities using that style, as well as others. For example, reading, making presentations, observing, doing—these are all different ways of going about learning, but not all of them may necessarily work for you.

ASSESSING YOUR PROGRESS

The final step in developing your learning contract is to identify ways to measure how well you are doing in meeting the goals. Terms like *measurement, evaluation, outcomes,* and *assessment* are often used to describe this step. What is important is that you are able to measure your growth objectively, using tangible activities or methods. For example, you could include in your contract such measurable activities as giving presentations that will be evaluated, creating resource lists that can be reviewed, keeping journals that can be read, developing written documents for review (Kiser, 2000; Rothman, 2000), or videotaping yourself, being observed and given feedback, or writing a case study for review by supervisors.

SPIRO It!

Another approach to developing the learning contract is to use a set of guidelines articulated in this acronym that emphasize the critical characteristics of specificity, performance, involvement, realism, and observability (Pfeiffer & Jones, 1972, cited in Royse, Dhooper, & Rompf, 2003).

Specificity refers to the specific rather than the general nature of the goals. A good one is "to learn how to assess for child abuse" rather than "to learn how to do the work of a human service worker."

Performance refers to the orientation of the learning goals, as evidenced in activities, duties, assignments, or responsibilities.

Involvement refers to the responsibility of the supervisor and the instructor to be involved in helping you achieve your learning goals. That involvement should be spelled out in the learning contract. For example, "Supervision will be conducted both individually (Tuesdays @ 3:00 P.M.) and in group format (Fridays @ noon)."

Realism refers to the need for goals to be realistic in nature. For example, to "supervise a caseload of five offenders" is a realistic goal for an intern with appropriate supervision; to "supervise a caseload of 150 offenders" is neither realistic nor appropriate.

Observability requires that feedback is available to the student and that it is provided through measurable, observable ways. Rothman's (2000) example cited earlier is an excellent example of the observability factor.

SUMMARY

You should expect some anxiety as you work your way through this early phase of your internship. We hope this chapter has helped you clarify some of your anxieties and identify their sources. We also have addressed the process of setting learning goals, activities, and measures, which should help alleviate some of the anxiety that comes from uncertainty. Many of the concerns discussed, though, have to do with the myriad of relationships you are beginning to form. In the next chapter, we turn to those relationships and the ways you can get the most from them.

For Further Reflection

PERSONAL REFLECTION:
SELECT THOSE QUESTIONS MOST MEANINGFUL TO YOU

1. Look at the "What ifs" listed at the beginning of the chapter. How many of them seem familiar to you? What can you add to the list?

2. Have you experienced feelings of fraudulence in your role as an intern? Do you ever feel like an imposter and fear being discovered as less competent than you should be and appear to be? Discuss your experiences with this phenomenon.

SPRINGBOARDS FOR DISCUSSION

1. This one involves your campus instructor, so be sure she or he is on board before you move ahead with the exercise. Make a list of both your anticipations—things that give you feelings of hope and excitement—and your anxieties—things that give you feelings of apprehension. Then send a copy of your list—anonymously— to your campus instructor so that your list and those of your peers can be copied and distributed in class for a discussion of similarities and differences.

2. As students approach the internship, they often use metaphors to describe what they think the experience will be like, such as a roller coaster ride, a glider flight, or a trust walk. Think of a metaphor that captures the way you are thinking, feeling, and behaving as you approach your internship. When sharing it with your peers, be sure to explain as well as describe it. Some of our favorites are *a long flight to a*

long-awaited destination; a roller coaster ride; stone soup (I am the soup, I am not the chef!); *planting a tomato garden*, and so on.

For Further Exploration

Horejsi, C. R., & Garthwait, C. L. (2002). *The social work practicum: A guide and workbook for students*. Boston: Allyn & Bacon.

Helpful and comprehensive discussion of approaches to learning contracts.

Kiser, P. M. (2000). *Getting the most out of your internship: Learning from experience*. Belmont, CA: Wadsworth/Thomson Learning.

Practical, informative discussion of one approach to learning contracts: learning goals, strategies, and methods of measurement, augmented by an exercise worksheet to develop a learning plan.

Rothman, J. C. (2000). *Stepping out into the field: A field work manual for social work students*. Boston: Allyn & Bacon.

Useful, informative discussion of one approach to learning contracts: learning goals, objectives, methods and approaches, and evaluation.

Royse, D., Dhooper, S. S., & Rompf, E. L. (2003). *Field instruction: A guide for social work students* (4th ed.). New York: Longman.

Pragmatic discussion of the application of the SPIRO model as a guide to developing a learning contract.

Wilson, S. J. (1981). *Field instruction: Techniques for supervisors*. New York: Free Press.

Informative, thorough coverage of various types of learning contracts and assessments.

References

Brust, P. L. (1986). Student burnout: The clinical instructor can spot it and manage it. *Clinical Management in Physical Therapy, 6*(3), 18–21.

Caine, B. (1994, Winter). What can I learn from doing gruntwork? *N.S.E.E. Quarterly,* 6–7, 22–23.

Caine, B. T. (Ed.). (1998). *Understanding organizations: A ClassPak workbook*. Nashville, TN: Vanderbilt University Copy Center.

Clance, P. R. (1985). *The imposter syndrome*. Atlanta: Peachtree Publishers.

Curtis, R. C. (2000, June). Using goal-setting strategies to enrich the practicum and internship experiences of beginning counselors. *Journal of Humanistic Counseling, 38*(4), 194–206.

Feldman, D. C., & Weitz, B. A. (1990). Summer interns: Factors contributing to positive developmental experiences. *Journal of Abnormal Behavior, 37,* 267–280.

Gordon, G. R., McBride, R., & Hage, H. H. (2001). *Criminal justice internships: Theory into practice* (4th ed.). Cincinnati, OH: Anderson Publishing.

Hodgson, P. (1999, September). Making internships well worth the work. *Techniques, 38,* 39.

Horejsi, C. R., & Garthwait, C. L. (2002). *The social work practicum: A guide and workbook for students.* Boston: Allyn & Bacon.

King, M. A. (1988). *Toward an understanding of the phenomenology of fulfillment in success.* Unpublished doctoral dissertation. University of Massachusetts, Amherst.

King, M. A., Spencer, R., & Tower, C. C. (1996). *Human services program manual.* Unpublished manuscript, Fitchburg State College.

Kiser, P. M. (2000). *Getting the most out of your internship: Learning from experience.* Belmont, CA: Wadsworth/Thomson Learning.

Nesbitt, S. (1993). The field experience: Identifying false assumptions. *The LINK (Newsletter of the National Organization for Human Service Education), 14*(3), 1–2.

Rogers, G., Collins, D., Barlow, C., & Grinnell, R. (2000). *Guide to the social work practicum.* Itasca, IL: F. E. Peacock Publishers.

Rothman, J. C. (2000). *Stepping out into the field: A field work manual for social work students.* Boston: Allyn & Bacon.

Royse, D., Dhooper, S. S., & Rompf, E. L. (2003). *Field instruction: A guide for social work students* (4th ed.). New York: Longman.

Sgroi, C. A., & Ryniker, M. (2002, Spring). Preparing for the real world: A prelude to a fieldwork experience. *Journal of Criminal Justice Education, 13*(1), 187–200.

Wilson, S. J. (1981). *Field instruction: Techniques for supervisors.* New York: Free Press.

Yager, G. G., & Beck, T. D. (1985). Beginning practicum: It only hurt until I laughed. *Counselor Education and Supervision, 25*(2), 149–157.

CHAPTER 6

Getting to Know the Clients

The interpersonal work is the critical aspect of any internship—
graduate or undergraduate. It is also the most difficult for
most people.
STUDENT REFLECTION

Regardless of how bright or motivated you are, or how much you know, it is your ability to form and maintain effective relationships with a wide range of people that will make you successful in your internship as well as in your career in human or public service. In the following chapters, we help you look at some of those important relationships—those with your clients (if you have them), your supervisor, and your coworkers—and examine your expectations and assumptions. We also call to your attention some issues you may not expect; they will be easier to deal with if they are not surprises. Finally, we deal with acceptance concerns, which are a normal part of the beginning of any relationship. We begin, in this chapter, with clients.

If you are working in a human service agency or other organization, it is easy to become preoccupied with paperwork, policies and procedures, and office politics. As an intern, you will have all these concerns to deal with plus the additional concern of being evaluated by at least two people. However, as J. Robert Russo reminds us in his book, *Serving and Surviving as a Human Service Worker* (1993), the people your organization helps are the reason for your job. If you are having your first experience with providing direct service, it is natural that many of your thoughts and much excitement and anxiety are directed at the clients. Even if your placement does not involve direct service, you may still want to skim this chapter in order to understand the broader context of your agency's work. We have found that concerns about clients generally fall into two categories: the nature of the client population (which involves your assumptions), and your relationship with clients. An area of emerging concern that is related to both these areas is personal safety.

THE NATURE OF THE POPULATION

As you begin your work, you no doubt have some expectations and assumptions about the clients you will meet. You probably have some knowledge of the clients as a group from the placement process and your first couple of days at the site. You know in general terms whom the agency does and does not serve and some of the needs the agency can and cannot meet. Beyond this factual information, though, it is important to be aware of the image you have of prospective clients. You are bound to have one, and it may not be entirely based on facts.

Imagine the clients for a moment. Each one will be somewhat different, but you can describe them within certain parameters. Client populations can be described as *homogeneous* or *heterogeneous* with regard to various characteristics. The more homogeneous a population is, the more alike its members are; the more heterogeneous the population, the more diversity is found within it. So, for example, a high school has a population that is relatively homogeneous in age, but may be heterogeneous with regard to characteristics such as race, social class, and intellectual acuity. Think about the clients' backgrounds, their reasons for coming to the agency, their personalities, life histories, and typical behaviors. Think about their race, ethnicity, social class, religion, and sexual orientation. How heterogeneous do you imagine the population to be? In our experience, interns often imagine that the clients will be very similar to one another, or they go to the other extreme and see each client as totally different, missing some of the important commonalities among them.

Now think about how similar and/or different the clients are from you. What do you have in common with them? What aspects of their lives and experience are totally unfamiliar? Here we often find that interns imagine they have almost nothing in common with their clients. That is, of course, not entirely true, but it is easy to feel that way. On the other hand, interns who have struggled with the same problem that brings clients to the agency (such as alcoholism or domestic violence) may assume that they know just how it is for the clients. A little thought, and some reflection on classes you have taken, will tell you that that's not true either, but it is another tempting assumption. Are you falling into either of these traps? (See Chapter 4 for a thorough discussion of this issue.)

You have been thinking about the image you have of the prospective clients. Now think about the source of that image. Probably very few of you have had extensive and varied experience with the client population you will be encountering. In spite, or perhaps because, of their inexperience, people make unconscious generalizations; they form stereotypes. The word *stereotype* has some pretty negative connotations, and you may be reluctant to consider that you hold any stereotypical assumptions. However, whenever you make a judgment about someone based on little or no information, you engage in stereotyping. Try imagining two people. One is a slim, slightly pale man in a tweed jacket, wearing wire-rimmed glasses and carrying a beat-up briefcase. The other is a tall, heavy man with a large belly, long hair, and a big beard, dressed in a T-shirt, sunglasses, dirty jeans, black boots, and a black leather vest with a Harley-Davidson insignia on it. In spite of yourself, are you making assumptions about what each of them does for a living? About how educated they are? Their personalities? Which one is more likely to have read existential philosophy? To have been in a barroom brawl? It seems to be a human tendency to generalize, and the fewer people in any group we actually know, the more we are likely to generalize from the few that we do know or have read about.

This example is based on physical appearance, but another way that interns—and others—get trapped into making assumptions is by becoming too focused on behavior. Some clients in human service agencies display pretty unusual or even bizarre behavior. It is easy to jump from there to all kinds of conclusions, especially if this is your first time working with these behaviors. Or you may find that, in spite of yourself, the reality of what some clients have done, especially those who have committed violent acts, makes it hard to see past that behavior to the unique features of each person.

Your assumptions can come from several places. Perhaps you have met a person with the same needs or heard one speak. Perhaps you have been such a client yourself or know someone who has. The media are another powerful source of images and assumptions. There are scholarly books and documentaries on various client groups, but a more powerful source of images is the mainstream media, especially television. Think, for example, about how many mentally challenged adults you have seen depicted on prime-time television. Were any of them leading anything like normal lives? How many of them had committed a crime? Couple the images of this population portrayed in the media with your own lack of direct experience, and you can begin to see where your image and stereotypes may originate. As one student said, "Given all the media attention to crime, in all the genres, it would be difficult not to form an impression of the criminal population."

As we said in Chapter 3, the first step in getting beyond stereotypes and assumptions is to admit you have them. Check with other interns; they have them, too. Theirs may not be the same as yours, but they have them. The next step is to gather as much factual information as you can, and hold your assumptions up to the light of objectivity.

At the beginning of this part of the chapter, we asked you several questions about your clients and asked what you were assuming. Now may be a good time to go back and try to find factual answers to those questions. Don't be discouraged if the answers don't come easily; you have held on to some of your assumptions for a long time.

Another common concern among interns is being successful with the client population. How would you define success with your client population? Be as specific as you can. What attitudes, skills, and knowledge do you think it takes to be successful with them? This may be a good time to talk with others at your placement site. Their input will help you obtain realistic answers to these questions as well.

ACCEPTANCE: THE FIRST STEP

As interns think about the kinds of relationships they want to have with clients, several concerns often arise. One of the most common is how clients will react to you (Baird,

2002). In talking with students who were going to perform community service with the homeless, Ostrow (1995) reported that one of their major anxieties was how they, as relatively affluent and fortunate young people, would be perceived by the homeless. You may be wondering, "What kind of reception am I going to get from these people? Will they respect me? Will they listen to me? Or will they just write me off?" The theme in these concerns is acceptance, which is a crucial part of the foundation for any relationship. You have probably read about how important it is for you to accept the clients, but they need to accept you, too. You need to find ways to have them accept you, and that's not always easy.

Think for a moment about what the word acceptance means to you. Don't recite a definition; try to put it in your own words. Ask a peer to do the same. You could probably have a debate about exactly what it means. Most people, though, know what it feels like when they are accepted and when they are not. Think about how acceptance *feels* to you as well as what it means.

Getting Clients to Accept You

Acceptance can look very different with different client populations. Some clients show their acceptance simply by being willing to talk to you; until they accept you, they just ignore you. Other client groups may show their acceptance by including you in their conversations, confiding in you, considering your suggestions, following your directions, or accepting the limits you set. How do you imagine you will know whether your clients have accepted you? Be as concrete and specific as you can.

Getting clients to accept you can be a real challenge. You will have an easier time with it if you can try to think about the situation from the perspective of the clients. They have probably had many frustrating and stressful experiences. In addition, they may be for the first time in a position where they cannot meet their own needs, where they have to let strangers into their business, and where their welfare is not in their hands (Royse, Dhooper, & Rompf, 2003). They may be expecting a miracle or a quick solution to their problems, and they are not going to find one. Or maybe they have just figured that out. If they are clients with a long history of services that did not help them, they may be expecting more of the same (Royse, Dhooper, & Rompf, 2003).

In some cases, clients may be particularly wary of interns. Some respond better to volunteers, reasoning that they are there because they want to be, as opposed to interns, who are fulfilling a requirement for school (Royse, Dhooper, & Rompf, 2003). In addition, interns generally do not stay with an organization very long. Some may be there for one academic year, and some may even be hired, but many are gone after one semester. Many clients have had unfortunate experiences with people leaving them or letting them down, and they are reluctant to invest in a relationship that will end soon. One of our students, who was working at a group home for adolescents, reported that one of the residents was quite hostile and rebuffed all attempts at contact. On further inquiry, the intern learned that this resident had gotten very close to an intern the prior semester and, although she knew the intern would leave, was devastated when it happened.

The main challenge is to enter, but not become lost in, the world of your clients. Schmidt (2002) writes of the power of perception in shaping our day-to-day experience, and the importance of entering the perceptual world of your clients. Entering that world means going beyond facts and figures, reports, and testing profiles, typically found in client files, and even beyond catalogs of experiences that clients may tell you about. It means trying to understand how their experiences have shaped the way clients see the world. In his book, *Crossing the Waters,* Daniel Robb writes about his experience working in a residential setting with troubled adolescent males (Robb, 2001). Here, he recounts an insight gained while working with one of the boys to fix a car: "I saw as we went that Louis had never had a man show him how to do anything. So, he expected to do it wrong, expected to be shamed for it, because unknown territory is not filled with angels" (p. 173). Here is another example from a student journal: "Most of these kids build a stone wall as a façade because it is the only way they have learned to survive. Behind that wall is an individual with feelings and emotions."

Part of their perceptual world, of course, is shaped by your clients' cultural profiles. These profiles must, to a large extent, be understood one client at a time, just as your own profile explored in Chapter 3 is different from almost anyone else's. Try to find out, then, not just what subgroups your client may belong to, but how that has shaped her or his behavior and perceptions. And try to avoid the trap of looking only for the weaknesses or liabilities in a client's cultural background. See the strengths as well. For example, a client whose ethnic background encourages him or her to seek help from family and friends, but never from strangers, may be reluctant to let you in. But remember that this same client has a wonderful resource in family and community, one that can be used to help. If you are a member of some dominant groups (see the definitions of *dominant* and *subordinate* in Chapter 3), clients from corresponding subordinate groups may be especially cautious around you. Finally, both your and your clients' stages of dominant or subordinate identity development will have an impact on the relationship. Many of you have probably studied this topic already, and this book is not the place for a lengthy discussion of multicultural issues. However, we provide some references for further reading at the end of the chapter.

Part of developing acceptance with clients is finding some common ground. A frequent concern for interns is their belief that they have nothing in common with their clients. A young woman interning in a shelter for single women and their children said:

> I don't have any kids. I have no idea what it's like to be a single parent, or to grow up with one parent. I have no idea how to survive on the streets or what it's like to be abused by a husband. What do I have to offer these women? Why should they listen to me?

Here is another thought from an intern working with adult drug addicts:

> I couldn't survive a minute in their world, and they know it. I didn't grow up in the city or in the street culture. I haven't been in a fight since grade school. And I'm going to give these guys advice?

These are real concerns, but remember that even having the exact same experience as your clients is not always a guarantee of success. Common experiences can help in establishing acceptance, but it can also be a hindrance. You may have had some of the same things happen to you as to your clients, but assuming or saying you know how your clients feel can be alienating for a client whose experience of the same problem was quite different from yours.

If you don't have much common experience, what can you do? You can't manufacture experience you haven't had, and pretending you know how it is will only make things worse. However, common ground does not always mean common experience, and showing respect for the clients' experiences is one way to gain their acceptance. Furthermore, even if you have not had the particular experience, you may be able to think of something in your background that gave you a hint to what the clients may be feeling. Again we turn to the words of Daniel Robb (2001, p. 56), who recalls being cut from a baseball team despite having superior skills:

> I turned and walked away from him, my good, dusty glove tucked under my arm, cursing the whole corrupt system. . . . It was clear to me at that moment that there was no trusting the men in charge. And I felt different. . . . I the kid who wasn't part of the team, the kid without a father. I had been cut. . . . Which all sounds maudlin and sappy until you remember the leverage of those days, the ferocity of your little peers, their cruelties and name callings. . . . The dangers of trust, the betrayals, the sweetness of being accepted and the tangles of rejection. These boys out here on the island, they were given no acceptance in the deeply psychological sense. . . . They're here because they were cut.

Another advantage to understanding the perceptual world of your clients is that you can try to understand their behavior in that context (Baird, 2002). This will help you not to be overly flattered by early compliments or expressions of trust. It will also help you not be overly flattened by clients who turn away from you and your attempts to connect. Remember that clients often have an agenda they are not going to tell you about, but one that you can understand once you come to know them. A client who approaches an intern and says, "You are the only decent person in this place—I can trust you" is either trying to gain an advantage or rushing into closeness on very little data, or perhaps both. Similarly, a client who gives you a hard time may be doing it precisely because you are new, to see whether you will back away, as so many others have done.

In counseling and therapy, this is called *transference*. Clients transfer to you feelings and reactions that are rooted in their past experiences. Marianne and Gerald Corey (2003) emphasize that transference can take many forms, including wanting you to be the parent, husband, friend, or partner they never had; making you into some sort of super-helper who can fix anything; refusing to accept boundaries that you set; and easily displacing anger or love onto you.

While you are working to enter the world of your clients, take care not to get caught in it. If their experience is similar to yours, it is easy to have old feelings come to the surface. It is also easy to become so absorbed in the experience of the clients that you take on the feelings and reactions that they have. This is called *overidentification*. Once that happens, as we discussed earlier, it is very easy to cross the line from wanting the clients to do well to *needing* them to do well. Also, you need to be able to help clients see the liabilities in the way they see the world, to confront them if necessary. Maintaining some distance, even as you get close to them, will help you here.

Accepting the Clients

Your ability to accept the clients is just as important to the relationship as their ability to accept you. In fact, it is more important. If you cannot accept the clients, they will know it in some way and on some level, and they will be far less likely to accept or trust you. Really, we are talking about a process of mutual acceptance here, and much of the work you did to get the clients to accept you will help you to accept them as well, especially the work you did to understand their perceptual world.

One important part of acceptance is to be able to see clients as more than their behaviors. Here are a few of the concerns that have been found in interns and beginning helpers, drawn from our experience and the work of Cherniss (1980), M. S. Corey and Corey (2003), and Russo (1993). How might you react to clients who:

- Lie to you?
- Manipulate you to get something they want but cannot have?
- Are never satisfied with what you have to give and always seem to need more?
- Become verbally abusive and physically threatening?
- Blame everyone else for their problems?
- Are sullen and give, at most, one-word answers or responses?
- Ask again and again for suggestions and then reject every one?
- Refuse to see their behavior as a problem?
- Make it clear they don't like you?
- Refuse to work with you?

Can you see these behaviors as understandable, although not helpful or praiseworthy, given what you know about the clients? Can you stay open to clients who do these things, and see their behavior as only one part of who they are? If not, take some time to consider why.

The most important thing you can do is try to understand your reactions. Remember, just as clients bring their past experiences and emotional tendencies to the relationship, so do you. *Countertransference* is the term counselors and therapists use to describe situations in which your perceptions of and reactions to clients are distorted by your own past experiences and hurts (Corey & Corey, 2003). Go back to Chapters 3 and 4 and reconsider some of the things you discovered or affirmed about yourself. Do any of those things help explain your reactions? For example, think about the reaction patterns you may have. If you have trouble confronting others, clients who challenge you overtly will be especially difficult for you. Knowing this will help you avoid assigning all the blame to the client. If your psychosocial identity, as discussed in Chapter 3, includes a shaky sense of competence, then clients who are not making progress may anger you more than they should. Finally, your cultural identity may help explain some of the reactions you are having as well. Consider this quote from a student journal: "Some clients evoked feelings of mistrust and prejudice, as well as feelings of sorrow. I am most shameful of the feelings of prejudice."

This quote shows an intern who is willing to look to herself, as well as to the client, to discover the source of this reaction. The willingness to admit prejudice is especially impressive. In time, this intern may move past feelings of shame and work on ways to overcome the prejudice and mistrust. Admitting these feelings, instead of pretending not to have them, is the first step.

Specific Issues

As you negotiate acceptance with your clients, be aware of a number of issues that seem to arise frequently as interns launch their relationships with clients. This awareness will help you prepare and not be shocked when they appear.

One particularly challenging issue concerns authority. In Chapter 4, we asked you to reconsider your feelings about being in a position of power or authority, which is perhaps a new experience for you. However, you have certainly been in a position where you were subject to authority, and so have the clients. Authority is often an important issue in forming a relationship with clients. Shulman (1983) has pointed out that clients tend to perceive you as an authority figure. This is especially true if you or your agency has legal authority over clients, such as the authority to deny benefits or to surrender an offender to the court. However, even when the authority is not explicit, you should be prepared for clients who see and react to you as an authority figure (Shulman, 1983).

Very often, clients are in a "one-down" position. They need something they cannot get for themselves. It could be something tangible, such as food or shelter, or something less concrete, such as control over a substance abuse problem or help in understanding some bureaucratic procedure. You and your agency have what they need; you are holding the cards. Think about when you have been in a similar situation. How did it feel? Clients will bring their fears about past experiences with authority to their relationship with you, and those factors will shape how they respond.

One way clients try to reduce their one-down feelings is to assess your background and experience, often with a goal of finding a flaw or an equalizer. We refer to this phenomenon as *credentialing*, and it takes place in nearly every internship. Sometimes clients will literally ask about your credentials, as when they ask about your education and training or experience with clients. Other times the credentialing is more subtle, such as, "Do you have children?" or "Have you ever been arrested?" It may be as simple as asking about your age. Sometimes clients are just trying to get to know you, but you may be surprised at the persistence with which these questions are asked. You may also be surprised by their reaction when they find the information they are looking for, and you may feel dismissed. Knowing that these assessments are a normal part of building a relationship will help you decide how to respond.

Clients will also often test your limits. They want to see where your personal boundaries are and whether and how you enforce agency rules. Because you are an intern, they may be genuinely unsure of your role and what you can offer them (Shulman, 1983). Some clients have experienced many workers, not to mention other people in authority, and have been treated in many different ways. They need to know what to expect from you, and although they may not like it when you set a limit, it ultimately helps them trust you. Other reasons for testing you may include a need for recognition and attention or an attempt to gain status with their peers (Shulman, 1983).

You may think of this phenomenon as occurring more with children, as they try repeatedly to pick up something they have been told to leave alone or try to poke or bite you. You may also associate such testing with people who are in an involuntary situation, such as juveniles in a detention center or runaway shelter, who refuse to do chores or curse in front of you to see what you are going to do. And in fact it is a very real issue in some criminal justice internships, where interns are often referred to as "officers in training" precisely so they are not seen as easy targets for limit testing. However, other kinds of clients can test you, too. An elderly client can press you to stay longer than you are able; a client at a soup kitchen may try to go through the line twice; a parent may ask you to stay late and watch the children until they can be picked up.

Interns are sometimes tested even more, because they are initially seen as "not real staff" (Gordon, McBride, & Hage, 2001). These behaviors can be exasperating, especially if you are not expecting them. Try not to imagine that really effective workers never have these challenges. Of course they do. They may have learned to handle them a bit more smoothly and quickly—and so will you, someday. Taking these challenges personally only makes it harder to meet them effectively. Remember that you once did this kind of testing, and you may still do it occasionally (Russo, 1993). Think back for a moment on your behavior with substitute teachers or baby-sitters. Perhaps you have even tested some of your college instructors in this way.

Another important issue in developing client relationships, and a frequent concern for interns, is self-disclosure. How much about yourself should you reveal to clients? There are actually two kinds of self-disclosure at issue. The first is personal information. It is natural for clients to want to get to know you, and they may ask you questions about your life and relationships. Some of these questions will seem quite comfortable and easy to answer. Others may not. You may feel that the questions are too personal or that they concern matters you would rather keep private. You may feel that certain information is inappropriate for certain clients. Given the earlier discussion of credentialing, you may also wonder about the motives behind the question.

These are issues you should think about, but not decide on your own. Your agency may have strict policies, and very good reasons for them, about how much you can disclose, and they may not always seem obvious to you. In some placements, interns are told not to have personal photos on display or divulge their last names. Matters of personal safety (which we will discuss later) and client boundaries sometimes dictate limits such as these. Be sure to discuss these issues with your supervisor.

The second kind of self-disclosure is more immediate. You will no doubt have opinions about and emotional reactions to things your clients say or events at the site and wonder whether it is appropriate to share your thoughts with clients. The question, "What should I share with clients?" is not possible to answer and, in fact, is not a helpful question. It fails to consider the wide variety of clients, situations, and goals that form the context of your work. It is better to ask a "three-dimensional" question (Hunt & Sullivan, 1974), such as, "What sort of self-disclosure is appropriate with which clients and for what purpose?" If you are working with a battered woman, for example, it may be appropriate to share some of your relationship struggles as a way of establishing an empathic connection; clients sometimes think their counselors have no problems of their own. However, if you are working with a heterosexual teen of the opposite sex, such disclosure is probably inappropriate, as it can blur an important boundary and create confusion about your intentions. Asking three-dimensional questions will help you, in collaboration with your supervisor, to find answers that work for you and your clients.

In Chapter 3, we stressed the importance of being aware of your values. In their book *Issues and Ethics in the Helping Professions*, Corey, Corey, and Callanan (2003) stress the importance of values awareness in relationships with clients. There will be times when you are dealing with a client whose values are very different from yours, and that can be perplexing and stressful for both of you. For example, suppose honesty and straightforwardness are strong values for you. Of course, you are dishonest occasionally, but you know it is wrong when you do it. However, you may be dealing with a client who has different values. In her experience, being honest, especially with human service workers and other authorities, means being taken advantage of by a service delivery system she perceives as unfair. For example, in some states, a woman on welfare will have her benefits reduced if she is married, even if her husband is not employed. So, she lies to you.

You may also encounter clients whose values about sexuality are different from yours. For example, you may have an adolescent client who is sexually promiscuous and thinks it is just fine. Furthermore, he uses no birth control and says a pregnancy would not be his responsibility. In the area of family values, you may be working with a client who thinks that a marriage must stay together at all costs, but you are in favor of divorce in some cases. Or you may be talking with a client who thinks that unmarried couples should not have children. One of the challenging aspects of situations like these is deciding how to respond. Corey et al. (2003) make a distinction between exposing and imposing your values on clients. You must decide whether to expose differences by telling clients about them. You must also decide whether to impose your values by attempting to influence clients to change theirs.

PERSONAL SAFETY

Increasingly we hear interns express concerns about their personal safety, and we are not alone. Berg-Weger and Birkenmaier (2000) cite several studies of real and perceived risk to social work interns. In the early days of their internships, our students often wonder whether the clients, the clients' families, or the neighborhoods where they work pose a risk of violence. In our experience it is easy to both overestimate and underestimate these sorts of risks. There are interns who worry when there is no realistic cause for concern, and there are interns who perceive no risk—and hence take no precautions—when the risk, while manageable, is very real. Whether you have these concerns or not, assessing levels of risk in your internship, learning to minimize risk, and learning to feel comfortable are all part of your task as you settle in with your client population.

How realistic are your fears? We cannot say. Certainly the potential for risk in human service work can be high. Kiser (2000) argues that society in general is an increasingly violent place, and that messages condoning violence are everywhere. Human services is such a broad field, though, that it is impossible to make general statements about risk. Here, however, are some examples, drawn from our experience, of internships with various levels of risk.

- Steven interns at a men's shelter in the city. It is the only shelter in the city that allows men who are actively abusing substances to be sheltered, and many men come in under the influence. Others congregate outside, seeking services only in a medical emergency. Furthermore, Steven is asked to ride along with an outreach worker, who travels to abandoned buildings to try and help addicts be more safe in their behavior, and perhaps to come for treatment.

- Carlos interns for an agency that monitors adjudicated youth in the community. He travels with a staff member all over the city, to schools, playgrounds, and apartments, to check on these adolescents and talk with them.

- Su-Je interns at a Student Assistance Center at a huge, urban high school. The school has students of every race, and from dozens of ethnic backgrounds, and intergroup tensions run high. Confrontations and fist fights are not uncommon. The center offers peer mediation, among other services, but sometimes these sessions erupt.

- Kavita interns at a women's shelter. She helps a staff member facilitate a support group for victims of domestic violence. One night, one of the clients tells the group that she is scared. Her abuser found out she was coming to the group and got very angry. He found her journal, and in it were names of clients and staff members. She sneaked out to group that night but is frightened about what he will do when he gets home and finds her gone.

Your level of risk depends on several variables, some of which are evident in the examples listed. Clients are one variable. Some clients are habitually violent; it is a response they have learned and are comfortable with. Some may become violent when under the influence of drugs. Others are in terribly frustrating and even desperate circumstances (Horejsi & Garthwait, 2002). If one of the adjudicated youths that Carlos is monitoring has gotten into more trouble and believes that he will be turned in, he may go to great lengths to stop that from happening. Non-clients are sometimes a risk as well. Parents whose children have been removed from the home, partners of victims of domestic violence, and friends of an adjudicated youth are all examples.

Location and hours of operation are another set of variables. Some human service agencies—and schools—are located in risky neighborhoods (i.e., at risk for violence or crime), especially for someone who does not live in that community. The neighborhoods that Steven travels to are more dangerous after dark. If your work takes you out into the community, as Carlos's and Steven's does, you may find yourself in such neighborhoods as well.

There are agency variables too. Some agencies have thorough procedures to help interns and staff stay safe and to handle risky situations. The high school where Su-Je interns gave her and all the staff careful instructions about what to do in case of a violent incident, when to respond, and how to get back-up. Others do not. Some agencies have conducted as complete an assessment of the risk factors involved in their work as they can; others have been less attentive.

You must confront your fears and assess the level of risk at your internship—even if your agency has not done that or not discussed it with you. Many interns are reluctant to discuss their concerns about safety, for fear of being seen as overly timid, not ready, or not committed to the field (Berg-Weger & Birkenmaier, 2000). That fear will keep you from confronting your prejudices, if you have them, and from making a realistic assessment of your situation. It may well be that some of your fears are unfounded, or exaggerated, because of prejudices and stereotypes. Some persons with mental illness or mental retardation exhibit some very odd behaviors, but that does not necessarily make them dangerous (Baird, 2002). A Caucasian intern in a predominantly Hispanic high school is not necessarily at greater risk for violence, but he may *feel* like he is. As we have said many times, the only way to deal with these stereotypes is to admit and confront them. If you are not comfortable discussing these issues with your supervisor, you may want to talk with your university instructor instead, or with your classmates.

You also need to be vigilant and assertive in gathering information about your placement (Berg-Weger & Birkenmaier, 2000; Kiser, 2000). Berg-Weger and Birkenmaier offer excellent suggestions, including finding out the number and nature of violent or abusive incidents that have occurred at your placement in the past year; getting a tour of the surrounding neighborhood; and gathering information about other neighborhoods you may be visiting or traveling through. You also need to understand any relevant policies and procedures for safety, crisis management, and reporting. Make sure you are trained in Universal Precautions for contact with bodily fluids such as blood or vomit. Finally, ask whether there is special training you can or should receive, such as cardiopulmonary resuscitation (CPR) or nonviolent restraint.

The likelihood is that once you have thought through your fears and gathered the information you need, you will feel ready to face what challenges there may be. If that is not true, if you feel like you are in over your head, you must discuss that with your university instructor as soon as possible.

SUMMARY

This chapter on clients may have raised more questions than answers for you. That is the point. We are trying to make you aware of and prepared for some of the challenges in the early stages of working with clients. We have provided you with some good material for reflection and for discussion with your peers, your instructor, and your supervisor. However, the exact shape and pace of your experience with clients are things we cannot know.

For Further Reflection

FOR PERSONAL REFLECTION:
SELECT THOSE QUESTIONS MOST MEANINGFUL TO YOU

1. What did you know about your clients when you started? What did you know about that population in general (e.g., juvenile delinquents or homeless)? Where did that knowledge come from? Did you have experience with the population before?

2. Now that you have gotten to know the clients a little bit, how have your initial impressions changed? Or have they remained the same?

3. How would you characterize the clients at your agency? In what ways are they similar to each other (race, gender, ethnicity, personality, social class, etc.)? In what ways are they different?

4. In what ways are your clients different from you? What do you have in common with them?

5. As you think about working with clients, what is most important to you about how they respond to you? About how they treat you? About how they feel about you?

6. Have the clients challenged your credentials? If so, how? If not, how do you think they might do so?

7. How easy do you think it will be for you to set limits with this population? What will be the issues around which you will need to set them? If the prospect seems difficult, how much of that difficulty is because of what you know about the clients and how much is because of what you know about yourself?

8. What aspects of yourself are you willing to share with the clients? What aspects are you unwilling to share and/or seem inappropriate to share with this particular population? How will you respond if you are asked about these areas?

9. What other kinds of challenges are you expecting from clients? Include general as well as specific challenges (i.e., those that pertain to one particular client). Why do you think these issues and behaviors will challenge you? Think about yourself. Is there something about you that makes these issues and behaviors especially troublesome?

10. If you have had other internships or field experiences, compare the current group of clients and your reactions to them with previous client groups.

11. What are some of the personal safety risk factors involved in your internship? Have you gathered the information you need? If not, where can you get it?

12. How do you feel about the level of risk you may be facing? If you are concerned, what can you do to feel more prepared?

SPRINGBOARDS FOR DISCUSSION

1. As a group, discuss the meaning of the word acceptance. Talk about what that word means to you, and how you know when you are and are not accepted. Then discuss what it might mean for the various client groups with whom you are interacting.

2. As a class, make a list of the populations that you are working with. Brainstorm a list of the more common stereotypes that society has about each group. Remember that you are not being asked whether you subscribe to these stereotypes, just what they are. Where do you think these stereotypes come from? You may want to revisit this list later in the internship to see how these stereotypes have held up to the light of experience.

For Further Exploration

Axelson, J. A. (1999). *Counseling and development in a multicultural society* (3rd ed.). Pacific Grove, CA: Brooks/Cole.

An excellent text on multicultural issues.

Berg-Weger, M., & Birkenmaier, J. (2000). *The practical companion for social work: Integrating class and field work.* Needham Heights, MA: Allyn & Bacon.

Thorough chapter on personal safety, with helpful exercises and inventories.

Corey, G., Corey, M. S., & Callanan, P. (2003). *Issues and ethics in the helping professions* (6th ed.). Pacific Grove, CA: Brooks/Cole.

Excellent section on the ethics of imposing values on clients.

Corey, M. S., & Corey, G. (2003). *Becoming a helper* (4th ed.). Pacific Grove, CA: Brooks/Cole.

Helpful sections on client issues.

Gordon, G. R., McBride, R. B., & Hage, H. H. (2001). *Criminal justice internships: Theory into practice* (4th ed.). Cincinnati, OH: Anderson Publishing

Excellent discussions of issues relating to clients. Obviously, this book is aimed at interns in a particular kind of setting, but it has many applications outside criminal justice.

Pedersen, P. (Ed.). (1985). *Handbook of cross cultural counseling and therapy.* Westport, CT: Greenwood Press.

Another good resource on multicultural issues. Triad model is helpful.

Pedersen, P. (1988). *A handbook for developing multicultural awareness.* Alexandria, VA: American Association for Counseling and Development.

A very "hands-on," short book with lots of exercises to help you.

Schutz, W. (1967). *Joy.* New York: Grove Press.

In his first stage of group development—inclusion—Schutz talks a great deal about acceptance concerns and the various ways of handling them.

Shulman, L. (1983). *Teaching the helping skills: A field instructor's guide.* Itasca, IL: F. E. Peacock.

Especially helpful regarding authority issues.

Sue, D. W. (1981). *Counseling the culturally different.* New York: Wiley.

Classic text by one of the leaders in the field.

References

Baird, B. N. (2002). *The internship, practicum and field placement handbook: A guide for the helping professions* (3rd ed.). Upper Saddle River, NJ: Prentice Hall.

Berg-Weger, M., & Birkenmaier, J. (2000). *The practical companion for social work: Integrating class and field work.* Needham Heights, MA: Allyn & Bacon.

Cherniss, C. (1980). *Professional burnout in human service organizations.* New York: Praeger.

Corey, G., Corey, M. S., & Callanan, P. (2003). *Issues and ethics in the helping professions* (6th ed.). Pacific Grove, CA: Brooks/Cole.

Corey, M. S., & Corey, G. (2003). *Becoming a helper* (4th ed.). Pacific Grove, CA: Brooks/Cole.

Gordon, G. R., McBride, R. B., & Hage, H. H. (2001). *Criminal justice internships: Theory into practice* (4th ed.). Cincinnati, OH: Anderson Publishing.

Horejsi, C. R., & Garthwait, C. L. (2002). *The social work practicum: A guide and workbook for students* (2nd ed.). Needham Heights, MA: Allyn & Bacon.

Hunt, D., & Sullivan, E. (1974). *Between psychology and education*. Hinsdale, IL: Dryden.

Kiser, P. M. (2000). *Getting the most from your human service internship: Learning from experience*. Belmont, CA: Wadsworth.

Ostrow, J. M. (1995). Self-consciousness and social position: On college students changing their minds about the homeless." Qualitive Sociology.

Robb, D. E. (2001). *Crossing the water: Eighteen months on an island working with troubled boys—A teacher's memoir*. New York: Simon & Schuster.

Rogers, G., Collins, D., Barlow, C. A., & Grinnell, R. M. (2000). *Guide to the social work practicum*. Itasca, IL: F. E. Peacock.

Royse, D., Dhooper, S. S., & Rompf, E. L. (2003). *Field instruction: A guide for social work students* (5th ed.). Boston: Allyn & Bacon.

Russo, J. R. (1993). *Serving and surviving as a human service worker* (2nd ed.). Prospect Heights, IL: Waveland Press.

Schmidt, J. J. (2002). *Intentional helping: A philosophy for proficient caring relationships*. Upper Saddle Falls, NJ: Merrill Prentice Hall.

Shulman, L. (1983). *Teaching the helping skills: A field instructor's guide*. Itasca, IL: F. E. Peacock.

CHAPTER *7*

Getting to Know Your Colleagues

As an intern, your clients, the community, or whatever the primary work of your agency may be are probably your biggest concerns and the major relationships on which you are focused. However, your colleagues can make or break your internship. We use the term *colleagues* to describe those people with whom you work—other interns, other professionals at the agency, and of course your supervisor. For many interns, the supervisor is the person whose opinion comes to matter most, who has the most impact in either a positive or negative way on their learning. Your faculty instructor is another key person in the internship process, and many of the issues you need to consider about your site supervisor also apply to your instructor. We won't be repeating all of them, but we will be reminding you to consider them. As for coworkers, you will no doubt work closely with some, and you may even work in teams. Whatever the case, you are bound to interact with them, and your relationships with your coworkers can have a substantial impact on your internship. In this chapter we focus on some initial concerns as you meet your supervisor and colleagues. Like any relationship, your relationship with your colleagues can initially feel like a good match or a poor one. And, like any relationship, it can be worked at, nurtured, cultivated, and improved.

SUPERVISORS

Certainly, one of the most important relationships at your internship is the one you have with your supervisor.[1] You will spend a great deal of time with this person, and he or she will be responsible at least in part for your evaluation. Your relationship with your supervisor can be a tremendous source of learning about the work, about yourself, and about the relationship itself (Bogo, 1993; Borders & Leddick, 1987).

It is also probably a unique relationship in your experience. You may well have had supervisors before, and maybe lots of them. But in a work situation the primary focus of supervision tends to be on performance. Performance is important at an internship, but the primary focus of your supervision is—or should be—educational (Royse, Dhooper, & Rompf, 2003).

Three definitions of *supervision* help capture the complexity of this role and relationship. Gordon, McBride, and Hage (2001) define it as "a process where two people … meet for the primary purpose of enhancing the personal and professional development of the student" (p. 32). Collins (1993), on the other hand, says that supervision is "a dynamic relationship of unequal power, which is characterized by intermittent closeness and distance" (p. 122). Still others see the role of supervisor as embracing both of these definitions. Moses and Hardin (1978) perhaps describe it best as going beyond instruction and developing skills and knowledge to promote self-actualization. Supervision should provide an extending experience that blends personal knowledge and personal qualities.

Your supervisors can and should help you learn a great deal. However, they also indisputably have power over you, and that has a real effect on the relationship. Interns report a variety of feelings about and relationships with supervisors. There is no one right way; in fact, a good relationship with a supervisor can take many forms. Consider these two examples from student journals:

> *I have strong convictions and so does my supervisor. We do not always see eye to eye. I admire people who can stand up for their convictions.*

> *My supervisor seems to know exactly what my needs are as an intern. She has never assigned me something I was not ready to do. I think the most important aspect of [an] internship is having good communication with the supervisor. And we have it.*

This section of the chapter will help you examine the beginnings of your unique relationship with your supervisors and consider some important aspects of that relationship.

Because the relationship is so important in so many ways, you would be unusual if you did not have some concerns as you approach it (Costa, 1994); probably your concerns focus on both task and process issues (Floyd, 2002). *Task concerns* center on the kind of work you will be given to do and how you will perform. You would not be unusual if you are worried about what will happen when you make a mistake and your

---

[1]In some cases, interns have more than one person functioning as an on-site supervisor. Your instructor on campus may also be fulfilling some of the functions of a supervisor. In this section, though, we are assuming that you have one on-site supervisor.

real and imagined shortcomings become visible (Baird, 2002). You may also be concerned about being assessed and evaluated. Your supervisors' perceptions of you are important in and of themselves, but the added prospect of a formal evaluation adds extra weight to those perceptions. Your own issues and personality will determine how important or intense any of these concerns are; and you may have some others we have not mentioned.

Process concerns center on your interpersonal relationship with your supervisor. Perhaps primary among those concerns is acceptance. Interns often wonder whether their supervisors will like them. More important, they wonder whether their supervisors will understand them and accept their weaknesses. As in any new relationship, they wonder whether they and their supervisor are going to get along well. Self-disclosure is a process concern in this relationship as well, although not in the same way as it is with clients. You may wonder how much about yourself you want to share or are expected to divulge and how many of your feelings and reactions you ought to reveal as you go through the internship. Since you do not yet know your supervisors very well, you may wonder how they will react to your disclosures; some interns, especially in counseling settings, worry that their supervisors will take the opportunity to analyze them and expose personal and professional weaknesses (Wilson, 1981).

Again, the cure for some of this is time. As you get to know your supervisor, you may or may not like what you find, but at least it will be known. And a trusting, comfortable relationship takes time to develop. However, clarity can also help, and that is the focus of this section. This is not a discussion about supervision theory. Rather, it is designed to help you become a more intelligent consumer of supervision. Looking at yourself as a consumer of supervision, with rights and responsibilities, is a much more empowering stance than looking at yourself as a passive recipient or even a victim. Being an intelligent consumer involves learning who your supervisor is as a person and a professional and examining the match between what you know about yourself and your needs with what your supervisor has to offer. It also involves clarifying some important procedures in the supervisory process, including the method and means through which you will be evaluated.

The Supervisor as Person and Professional

In our experience, interns often bring certain preconceptions about supervisors to the internship (see also McClam & Puckett, 1991). They also bring expectations, hopes, and desires. They expect their supervisors to function in a variety of roles, and to do them all well (Alle-Corliss & Alle-Corliss, 1998; Baird, 2002; Borders & Leddick, 1987; Kiser, 2000; McCarthy, DeBell, Kanuha, & McLeod, 1988; Royse et al., 2003). Many interns hope that, as a *teacher*, their supervisor will help them set goals and objectives, help them learn new skills, and be sensitive to their particular learning styles. They hope that their supervisors will be *role models* with great stores of experience and expertise. They also hope that supervisors will use their counseling skills to be a *support* to them and help them through difficult emotional times and decisions. In the role of *consultant*, interns want their supervisors to help assess a problem or situation and generate alternative courses of action. Supervisors are in the role of *sponsor, advocate, or broker* when they take an active interest in the career of an intern, both at the placement and beyond (Kanter, 1977; Speizer, 1981). And some interns hope that their supervisor will become a *mentor*, someone who takes a deep and special interest in them as a person and a budding professional.

The reality, of course, is that no one can do all of these things equally well, and that your supervisor is a person with concerns, foibles, strengths, and weaknesses just like you. Being an intern supervisor is a complex task, for which many supervisors receive little or no formal training (Baird, 2002). They may have a lot of experience as a supervisor, but everyone has to have their first intern sometime, and that intern may be you. Many supervisors agree to take on an intern because they see it as a way to contribute to their profession or perhaps to return the favor that a supervisor did for them when they were students. There are some more tangible benefits to working with interns as well. For example, some supervisors have sought out interns because of the positive effects the supervisors observed on the morale, commitment, and proficiency of their staff as a result of working with interns (King & Peterson, 1997).

Your supervisor also has roles as a worker and perhaps as an administrator or a supervisor of agency personnel, in addition to the role of intern supervisor. Your supervisor also probably has a supervisor as well, or maybe several of them. All of these roles place demands on your supervisor's time. Agreeing to supervise an intern is yet another investment of time (Baird, 2002; Royse et al., 2003). Sometimes your supervisor has been released from some other responsibilities in order to supervise you, but not always. If the investment pays off and the intern does a good job, then everyone benefits, and the intern may even be able to take some of the workload off the supervisor. If it is a difficult internship, on the other hand, and it requires a great deal more time and energy than anticipated, the return for the agency may not be as great. These concerns are probably on the mind of your supervisor. And remember that out of all these roles and responsibilities, your supervisor's top priorities are and should be the clients and communities that the agency serves.

Try not to make assumptions about your supervisor's level of experience, which may not be a great deal more than yours. Remember, though, that experience is not always the best indicator of quality in a supervisor. Just as brilliant scholars may not make good teachers, because students' struggles with the material are incomprehensible to them, so it is that experts in your field may not be the best supervisors. They may have forgotten what it is like to be new to the field, and the gap between your skill levels may be so great that they cannot function as effective role models for you—the expertise seems unattainable.

So, a less experienced person is not necessarily a problem. Regardless of her or his level of expertise, your supervisor has a different, more objective vantage point from which to view your experience and your struggles.

Supervisory Style

In all the roles we mentioned your supervisor has a unique style. Just as learning style is multifaceted, so is supervisory style; many different dimensions are involved. Some aspects of your supervisor's style may be the result of careful study and deliberate choice. Others are the result of attempts to replicate the way she learned best or to emulate a favorite supervisor. Still others are matters of personal preference. Some aspects of a supervisor's style may not even be deliberate; he may be quite unaware of it or assume that everyone does things in the same way.

No one style of supervision is good for everybody. You may have heard great things about your supervisor, yet be puzzled when the relationship doesn't seem to be working. It may be that you are a person who is better suited to a different style. Furthermore, many authors have discussed a developmental approach to supervision, suggesting that the needs of supervisees vary according to the stage they are in (Galassi & Trent, 1987; Hersey, Blanchard, & Johnson, 2001; Inkster & Ross, 1998; Loganbill, Hardy, & Delworth, 1982). An important step in developing a relationship with your supervisor is to learn more about some of the key components of your supervisor's style.

The work of an individual, and sometimes of an entire organization, is usually guided by a particular theoretical orientation to the work. There are, for example, many different theories about teaching, counseling, management, and community organizing. It is important that you know what theory or theories are the most important to your supervisor, and how tolerant of other styles your supervisor is. If, for example, you favor a humanistic or psychodynamic approach to counseling and your supervisor believes in a cognitive behavioral approach, you will need to know to what extent you are expected to use that style with your clients. Or, if you tend to focus on systems change and community advocacy and your supervisor focuses on individual work and adjustment, you will need to have a serious conversation about how best to work together.

The terms *expressive* and *instrumental* describe two general approaches to management or supervision (Russo, 1993). Supervisors who use expressive approaches are more people oriented; their primary concern is for people and relationships. Supervisors who use this style often need to be liked and appreciated, and they cultivate friendships with those they supervise. A supervisor with a more instrumental orientation, on the other hand, is primarily concerned with productivity and task accomplishment. Please note that this is perfectly possible even when the business of the organization is people. A supervisor at a social services agency can be concerned primarily with serving the maximum number of clients and achieving measurable results. For these supervisors, respect is important in the supervisory relationship, but not necessarily affection and friendship. They are able to handle conflict and hostility from supervisees effectively.

Corey and Corey (2003) discuss a related issue when they compare supervisors who use a more confrontational approach with those who are more supportive. A confrontation is not necessarily an angry encounter. A confrontational supervisor will spend less time offering support and more time helping you see what you could have done differently. For example, if you have just had a difficult encounter with a client, your supervisor could spend time listening to you talk about how that felt and offer some reassurance and empathy. On the other hand, she or he may ask a series of pointed questions designed to help you see where you misjudged the situation and/or used an ineffective intervention. This latter technique does not necessarily signify a lack of caring about you; it is merely a different way of caring and a different perception of what you need.

Hersey and Blanchard (Hersey, Blanchard, & Johnson, 2001) have identified two dimensions of supervision, which can be combined in four different ways. *Direction* is the first dimension, which involves giving clear, specific directions, close supervision, and frequent feedback. *Support*, the other dimension, is a nondirective approach marked by listening, dialogue, and high levels of emotional support. So, if your supervisor tells you to write a report and says, "There are lots of good ways to do this, and I know you are nervous, but you can handle it. Please ask all the questions you need to and let me know when you are done," this is a high level of support and a low level of direction. A directive approach might tell you exactly what the report should look like and what it should contain, with frequent deadlines for the submission of sections and the final report. These two dimensions can be combined in four ways: high support/low direction; high support/high direction; low support/high direction; and low support/low direction.

You may think that high support and high direction would be the best all the time. Of course, it is frightening to be given a task and have no idea how it should be done. However, too much direction over a long period of time can discourage independence and deny you the opportunities to try something and learn from it. Sometimes you need to take risks, and even make mistakes, and that is difficult when you are very closely supervised. Similarly, everyone needs some support and encouragement, but too much can keep you from seeing where you need to improve. It can also make you dependent on your supervisor in a different way; there may be times when you need to function without considerable external support.

Hersey and Blanchard (Hersey, Blanchard, & Johnson, 2001) suggest the most appropriate combination of support and direction depends on the maturity level of the supervisee, and they use the term *maturity* in a particular way. First of all, the term is meant to apply to a specific task or set of tasks that the supervisee is facing; it is not a global description or a personality trait. They have identified four levels of maturity, which they describe as a combination of willingness and ability. Each of those levels is best suited to a particular combination of support and direction. For example, suppose an intern is going to be asked to lead a support group. It may be that the intern, at least at first, is neither willing nor able to do it. In that case, the supervisor provides a lot of direction about how to lead the group, and monitors the progress carefully. At the second level of maturity, the intern may feel unable, but willing to try. In this case the intern needs high levels of both direction and support. On the other hand, if the intern has the skills but feels uncertain (able but not willing) then high support and low direction is needed. Finally an intern who has the skills and the willingness does not need much support or direction with regard to the task of leading groups. In other areas and for other tasks, the maturity level and needs may be quite different for the same intern.

Another dimension of supervisory style to consider is collaborative versus hierarchical approaches (Costa, 1994). A *collaborative* approach emphasizes mutual dialogue between supervisor and supervisee and encourages divergent thinking. In a *hierarchical* approach, however, the supervisor serves as a communicator of expert knowledge. Expertise flows in one direction, as do the questions, except when the supervisee needs clarification on something.

Horejsi and Garthwait (2002) have identified some additional dimensions of supervisory style, including supervisors who like things structured and routinized versus those who are more fluid and flexible; those who focus on details versus those who focus only on the larger picture; and those who like a fast pace versus those who are more "laid back." Keep in mind that each of the terms used here to describe supervisory style defines one end of a continuum. There is plenty of room in the middle. As you get to know your supervisor, you can locate her or him on each of these continua. Remember, though, that supervision is a relationship, and so far we have only covered one side.

Your Style

As you read through the preceding paragraphs, you probably found yourself identifying your own preferences among the styles discussed. If not, take time now to review the section. As you do, note your reactions and preferences, and remember that they are influenced by a variety of factors.

Your learning style was discussed in Chapter 3. You may learn best by reading, by observation, or by direct experience. Your supervisor may have a different style, assigning readings, for example, when you are longing to deal with real people and situations. Another factor to consider is your motivational style (Borders & Leddick, 1987). Some people respond best to warmth and praise but wilt under blunt critiques, whereas others may become impatient with too much warmth and praise and prefer a relatively impersonal, objective critique.

In Chapters 3 and 4, we encouraged you to consider some of your issues and patterns. It is especially important that you consider any patterns or unfinished business you may have with issues of *authority*. We already asked you to think about the challenge of being in authority. Here we are asking about negative experiences or reactions you have had to people with authority over you. Some people have a hard time accepting authority. Some go to the other extreme and have difficulty challenging it, even when a challenge is warranted. Sometimes older students have a separate set of challenges in supervision (Royse et al., 2003). Perhaps they have managerial experience in another field. Perhaps they spent a long time rising through the ranks to get there. If so, it may be hard for them to accept being in a subordinate position in supervision, especially if there are some aspects of their supervision that they do not like or that leave them feeling disempowered. Your psychosocial identity may also be important to remember in your relationship with your supervisor. For example, your sense of competence—one central component of that identity—may be challenged by supervision, and your sense of initiative will be important when you are given projects of your own.

Another dimension of your style is as a separate or connected knower. This style may lead you to prefer abstract analysis over in-depth case study, or vice versa. Twohey and Volker (1993) have pointed out that both feelings and principles have a place in supervision. If you are having a difficult time with a client, for example, it is important to examine the theoretical and ethical principles that are relevant to the case. However, you may have feelings that are not easily reduced to a theoretical quandary or a clash of ethical principles. You may be more concerned about the client and the potential disruption of your relationship with him or her. This aspect of your experience needs to be attended to as well.

Matches and Mismatches

If you—or your supervisor—are expecting a certain kind of response and get a very different one, it can be disconcerting. If you are looking for direction and get support, or expect collaboration and get hierarchy, for example, it can be confusing and upsetting. A supervisor expecting a theoretical analysis who instead gets a long discussion of your emotional reactions may feel that you either did not understand or are not capable of answering the question.

These moments are never easy, but they are easier if they are understood as a mismatch of styles. Your supervisor is not necessarily insensitive to your needs any more than you are stubborn or inept. You may each be doing exactly what you think is appropriate and required. Misunderstandings can be avoided by taking a proactive stance toward your supervision.

Most supervisors do not mind being asked about their style. They may or may not be familiar with the specific theories discussed here, but you can find a way to ask about these aspects of style without using the particular jargon described in this chapter. Your supervisor may also be familiar with and use other theories about supervisory styles. If so, you can learn both about that theory and about your supervisor when discussing style. You may also want to consider discussing your style with your supervisor, including your preferences and needs. These needs do not have to be demands. You are merely considering your style, along with your supervisor's style, and looking for potential areas of match and mismatch.

Structural Issues in Supervision

Part of what creates anxiety for anyone is the unknown. Each supervisor and placement site is different, of course, but in our experience, there are some common features of the supervisory process, and you will want to be clear with your supervisor about each of them.

GATHERING INFORMATION

Supervisors need information about interns in order to know how they are progressing and developing. This information guides supervisors in their conferences with you and in the evaluation process. There are several ways for supervisors to gather this information. The approaches all have strengths and weaknesses, and your supervisor may use more than one of them. It should be fairly easy for you to find out which of these approaches will be used.

One common approach is *live supervision*, which means that your supervisor either works alongside you or observes as you work. In some settings the observation can be done behind a one-way mirror, so that neither you nor your clients can see the supervisor, although you know you are being observed (and the client may be told as well). There is no substitute for live supervision; it is the only way the supervisor can see your work firsthand. Your nonverbal behavior and nuances of tone of voice are things that you will not notice, but your supervisor can (Baird, 2002). However, live supervision is often not possible. It can be intrusive on the relationships with clients. It can also make interns nervous, and more important, having your supervisor right there and available for consultation may encourage you to be overly dependent on that help.

Another somewhat "live" approach is the use of *audio-* or *videotaping*. You make a tape of yourself and go over it with your supervisor. Some supervisors will go over the tapes privately first, and some may ask you to prepare an analysis of the tape as well. While no one we know (including us) enjoys being taped, it can be of enormous value. Watching or listening to yourself is an excellent way to notice voice patterns, body language, patterns of movements, and other aspects of your performance that you would never be aware of otherwise.

Self-report is another common approach to data collection. Your supervisor will ask you to report, orally or in writing, on your progress with clients or projects. Some agencies use a highly structured and prescribed format for these reports, and others are more flexible. Self-report encourages you to reflect on your own performance and to take responsibility for critiquing your own work, which is something you are going to have to learn to do eventually. It is also a chance to sharpen your analytical skills. However, it will never uncover the kind of unconscious patterns and blind spots that live supervision and taping can reveal. There is also a natural human tendency to describe yourself in the best possible light, especially when you want to make a good impression (Borders & Leddick, 1987).

Peer report is a less common approach. In cases where your supervisor cannot observe you directly but other staff members can, the supervisor may request written or verbal reports from these staff members. In the case of written reports, you may or may not be able to see them. It is important to discuss your placement's policy on gathering information on your performance as early as you can.

CONFERENCES

Kiser (2000) points out that you will have lots of different interactions with your supervisor, ranging from quick conversations in the hall to long talks as you ride to an appointment together. None of these, however, constitutes formal supervision. You should expect to have regular meetings with your supervisor throughout the internship. This is a time for you to report on your progress, ask and answer questions, and get feedback. Lots of supervisors will also meet with you spontaneously if there is something important to discuss, and that is fine. However, those meetings allow you no time to prepare, reflect on your experience, and think about issues and questions you may have (Royse et al., 2003). We strongly recommend that some of your meetings be at regularly specified times. Find out from your supervisor how frequent and how long these meetings will be and schedule them ahead of time. It is easy, in the crush of crises and other responsibilities, for interns and supervisors to be tempted to cancel supervision, especially if things are going well (Baird, 2002). Don't do it—few things are more important to your learning than regular and productive supervision.

Another important consideration is the structure of the meetings (Baird, 2002; Borders & Leddick, 1987; Horejsi & Garthwait, 2002; Leddick & Dye, 1987). Some supervisors use a highly structured format. They ask the questions, and they ask at least some of the same ones each time. They may also have a structured way that they want you to present your cases or your progress on projects. Still others emphasize more didactic approaches, where they do most of the talking, teaching you about different theories, asking you questions about readings they have given you, ands so on. Other supervisors are very loose in structuring the session and respond only to what comes up in the conversation. Some supervisors come into conferences with a set agenda. Others let you ask the questions and raise the issues. Some, of course, do both.

Your supervisor's choice of structure and activities may be based on personal preference, deep beliefs, or assumptions about what will be most helpful to you. Some of
them are more suited to particular learning styles than others. So while you should stay
open to all forms of supervision, spend some time thinking about how well suited each
of these approaches is to the way you learn best.

Regardless of the structure, it is important that you reserve some time to plan for
these conferences (Berg-Weger & Birkenmaier, 2000; Kiser, 2000; Royse et al., 2003).
Remember the learning cycle described in Chapter 3. At most placements, there is so
much to do that there is a temptation just to work, work, and work some more. Conferences force time for reflection, which is a critical part of completing the cycle, and
preparing for them does as well. Planning for conferences can also empower you and
give you more control and influence over the session. You should set aside time to
think about questions, observations, and problems you want to discuss, and to prepare
supporting documents if necessary. You may need to put them in some kind of priority
order, considering their importance to you and their immediacy in the lives of clients
or the completion of projects (Kiser, 2000). You should also be ready to report on your
progress with clients and projects.

Your Reaction to Supervision

Even under the best of circumstances, the process of supervision often produces at least
some anxiety. In your internship, you are being asked to try out new skills and approaches. In your supervision conferences, you are often encouraged to discuss your
feelings and reactions. A supervisor who feels you are running into some unresolved
personal issues is probably going to let you know that. Under those circumstances, most
of us experience two seemingly contradictory emotions. On the one hand, we are there
to learn and grow, and supervision is a tremendous opportunity for that. On the other
hand, we do not always want to see ourselves clearly or change our ways of doing things.
We want to learn and change, yet we resist learning and changing (Borders & Leddick,
1987). Some of us are even resistant to compliments, as in this student's journal entry,
"Positive feedback is something I have a hard time accepting. In fact, I almost cry when
I hear it. I am learning that my work in the world is important and well received."

These conflicting emotions sometimes manifest themselves in behaviors that help
you avoid really looking at yourself and your progress (Borders & Leddick, 1987; Costa,
1994). They include being overly enthusiastic, avoiding certain topics, and going off on
tangents. You may also become forgetful, especially about projects, assignments, or issues that make you nervous. Some interns become argumentative, taking issue with
every point the supervisor makes. They may or may not do so out loud, but privately
and to their friends they dispute every criticism. Others go to the opposite extreme and
agree with the supervisor even when that is not how they feel. If you find yourself ex-
hibiting any of these behaviors, think about the feelings and concerns that may be behind them. The point here is not that any single instance of one of these behaviors
indicates that you are avoiding an issue; however, when they become patterns, you
should look very carefully at them, discuss them with your instructor, and find a way to
work through them.

The Evaluation Process

At some point in your internship, you will probably have a formal evaluation. At this time, the supervisor lets you know how you are doing and have done overall. In our experience, most internship programs request two evaluations: one at the midpoint and one at the end. It is natural to be apprehensive about the prospect of reading or hearing an evaluation. But try to remember that the evaluation is another opportunity for growth, empowerment, dialogue, and even assertiveness. Above all, it is an opportunity to learn. Again, the more you know about the process, the less mysterious and threatening it will be, and the more you can direct your emotional energy at making the most of these opportunities. As one intern said after reflecting on her evaluation, "He made me curious about myself and want to work on the areas that need improvement."

You can take the initiative to find out several things about how you will be evaluated. The structure of your evaluation is an obvious concern. Find out how often you will be evaluated and whether it will be oral, written, or both. Gather information on the specific format. Some written evaluations use scales, where you are given a number from 1 to 5 or 1 to 10, to indicate how well you have met various criteria. Some evaluations let the supervisor write in a more narrative form. Still others use both formats.

It is also important to know what standards are being used in your evaluation (Wilson, 1981). Some supervisors will compare you to other interns, either current ones or those from past semesters. If you are among the best, you receive high ratings. Others evaluate interns individually, assigning ratings based on how much they have grown over the course of the semester. Still others have a standard set of expectations and criteria that you must meet to get a high rating. If the supervisor is using numerical ratings, find out what they mean. What do you have to accomplish to earn a rating of 3?

We knew one student whose supervisor had nothing but praise for her at their weekly conferences. Yet the midsemester evaluation contained ratings no higher than 3 on a scale of 1 to 5. The supervisor explained that he never gives anyone a 5, because no one is perfect, and if he gave 4s at midsemester, there would be no room for improvement. To him, the 3s were an excellent midsemester evaluation, but to the intern, they felt like an average rating, much like receiving a C on a paper.

You will also want to know the function of the evaluations. Who can see them and for what purpose? Are the midsemester evaluations recorded and counted as part of your grade? What portion of your internship grade is determined by the supervisor's evaluation? You should also clarify when and under what circumstances you will see your evaluation. Many supervisors will make the written evaluation part of a conference. They may discuss the evaluation before filling out any forms, or they may go over the written evaluation with you. Regardless of the specific arrangements, you should see the evaluation before it goes to anyone else and have an opportunity to review it and ask questions.

Your Faculty Instructor

The role of instructors in an internship varies from school to school, but at the very least this is the person who will be grading you on your overall performance. If the instructor is also leading a seminar, reading and reacting to your journal, or visiting you on-site, then in some ways that person is supervising you, too. And while the relation-ship is generally less intense and the contact less frequent than with your site supervi-sor, many of the same concerns and issues can and do arise. Acceptance is a concern in this relationship, as is assessment. It is important to get a sense of the teaching, and per-haps the supervision style of your instructor and see how they match with your needs. Review the section on supervision with your instructor in mind.

COWORKERS

Although your supervisor may be your most important colleague at the placement site, you may actually spend more time with other staff members. Sometimes interns report that a particular staff member becomes an informal supervisor and that they learn more from and feel more supported by that person than their assigned supervisor, whom they may only see once a week. Whatever the relative importance of coworkers to you, they are a great opportunity for learning and a potential source of support. They may be mentors, sponsors, and role models offering you knowledge and skills, an in-sider's and veteran's perspective on the agency and its work, and valuable feedback on your performance.

Just as with your clients and supervisor, you probably have an image of your coworkers, and it needs to be tested. And just as with clients and supervisors, another major concern of the Anticipation stage is acceptance from coworkers.

Expectations

You may not have thought much about it, but you probably have an image of what your coworkers will be like. What level of education do you think is needed for work-ing at your site? What level of education do you think most of the staff has? Earlier, we asked you what attitudes, skills, and knowledge were necessary to work successfully with the clients at your site. If you skipped the section on clients, take a moment now and think about what it takes to be successful at your site, not as an intern but as a full-time professional.

Most people are guided by some standards or principles in their work. Your agency has a set of rules and policies and may even have a code of ethical behavior. You may have studied ethics and other aspects of professional behavior in your classes as well. To what extent do you think your coworkers are aware of these issues and standards? To what extent do you think they adhere to them? What do you imagine happens to them if they violate those standards?

We ask these questions because in our experience interns encounter all sorts of coworkers, some of whom behave very differently than they were expecting, as the fol-lowing student journal entries make clear:

> *During recreation period, I am out there trying to make contact with the kids, and the other staff members just sit and talk with each other—and they're the ones getting paid!*

In private, they make jokes about the residents. They have derogatory labels for almost every one of them.

It seems that my main value to some of them is as a gofer. They give me jobs they don't want, so they can take a break.

I was amazed to find out that several of the staff here are brand new in the field.

With my previous field experiences and practicum, I have more experience than they do.

Some of them don't have degrees in human services or psychology. Some of them don't have degrees at all!

You can, and should, spend some time getting accurate information about your coworkers. Find out what the hiring process is at your placement. What level of education is in fact required? What other requirements and preferences does the agency have? How many staff people leave the agency in an average year? What do people have to do to get promoted? You may be quite surprised by the answers to these questions.

Acceptance

Most interns are concerned about whether, when, how, and on what terms they will be accepted by the staff at their placement site. As you ponder this question, there are a number of issues to consider. Obviously, different individuals will react to you differently.

Based on what you know now, who are the people at your site whose acceptance is most important to you? Remember, too, that acceptance is not the same as being liked. Most of us want to be liked, but the staff members do not have to like you in order to accept you. If you need every staff person to like you, you may find yourself taking action directed primarily at achieving that goal rather than at what is best for your agency or your clients. On the other hand, here is a reasonable, although not always achievable, desire expressed by a student: "What I want from them is to remember when they first started out and appreciate what I am going through."

Remember, too, that acceptance is a two-way street. One challenge for you is to accept your coworkers for who and what they are. Again, you do not have to like all of them or even respect everything they do. In your classes, you may have studied and practiced how to be accepting of clients whom you may not like or approve of. Your coworkers are not clients, and they are accountable to standards that your clients are not, but they deserve the same accepting, nonjudgmental treatment that you extend to your clients.

Depending on the size of the organization hosting your placement, you will probably get a range of reactions from the staff, from warmth to indifference and even to hostility. Some interns report feeling like guests, some like employees, and still others like intruders. These two quotes from student journals are a study in contrast:

I felt like an outsider at first. I didn't know anyone at lunch and they would all sit together.

They are very helpful in that they take time to explain what they are doing and why. They don't seem to be annoyed with my questions and inexperience.

Before you are too harsh on the staff members mentioned in the first quote, try putting yourself in their place. It is another busy day at work, and they are meeting an intern whom they may or may not have known is coming. They may be wondering what your presence means for them, what you want from them, exactly why you are there, and whether they can count on you. The staff's reactions to you depend on many factors, including their personalities, past experiences with interns, understanding of what an intern is (or lack thereof), and their relationship with your supervisor (Gordon, McBride, & Hage, 2001). Some of them may even resent you. Energetic new workers can be threatening to those who are disenchanted with or exhausted by their work. A new intern's ideas—or even willingness to follow established procedures to the letter—can be threatening, or some staff members may fear for their jobs.

After the visit, I began making notes on my clipboard. The worker I was partnered with asked me what I was doing and when I told her she said, "Oh. We don't have to do that." I said that my training manual says that we do, and she just shrugged. Later on I asked my supervisor, without mentioning what she said. I was right. Later I found out that she was complaining and making remarks about me behind my back.

Furthermore, agencies sometimes use interns and volunteers as a way to stretch their staff and meet their needs without hiring additional people (Suelzle & Borzak, 1981). This practice may violate labor laws or ethical policies and often results in a lack of continuity within the agency, as well as resentment from front-line staff members who have to deal with the results.

Even staff members who have no experience with interns may have heard stories about interns from your school or other schools. As we mentioned earlier, when an intern works out well, it is productive and exciting for everyone involved, but if not, it can be very taxing on staff time and energy and can be disruptive to the work.

Gordon, McBride, and Hage (2001) have noticed that interns often struggle with feelings of intrusion and marginality. If you feel like you really don't belong at the placement, that you don't fit in and can't contribute much, you are struggling with feeling like an intruder. On the other hand, if you are feeling like the staff doesn't want you there and doesn't see that you can be of much help, then you are struggling with marginality. Of course, as the following student journal entries show, both can occur at once:

I feel like my co-workers treat me like an adolescent if I don't present myself properly.

They tell inside jokes and don't clue me in. One of them in particular just looks at me, as if I don't belong there and she is waiting for me to prove it.

In fact, you may feel like an intruder at first, and you may be treated like one. It may take some time to prove yourself to the staff. Over time, these feelings and reactions usually dissipate, but they are troubling while they are present.

You may also find yourself surprised by the variety of approaches to the job taken by your colleagues. Some may be very different than what you expected. You may be disappointed in them, and you may have a right to be. However, understanding a little bit about what can happen to a person over time may help you understand what you are seeing.

BURNOUT

First of all, it is possible that some staff members are experiencing burnout. Burnout is the result of persistent job-related stress. Its effects include physical and emotional exhaustion, rendering the worker depleted and drained; depersonalization of work and clients, so that the worker feels detached and even callous; and reduced personal accomplishment (Maslach, 1982). The exhaustion is not just physical; it is emotional and even spiritual. The person feels irritable and displays negative attitudes toward clients. Productivity declines, and the person often feels isolated and withdrawn. Perhaps you have had firsthand experience with burnout, encountering it in teachers, school counselors, or staff at other field sites.

You will find people who are burned out in every profession, but human service workers seem especially vulnerable. The kinds of people who select human services as a career are typically concerned with individuals and their problems, attuned to human suffering, and anxious to make a difference. These very qualities make them vulnerable to working too many extra hours or to putting their clients' needs ahead of their own (Cherniss, 1980; Corey & Corey, 2003; Schram & Mandell, 2002). There are also many features of human service work that can cause stress. Clients are often not appreciative of your efforts, change often comes slowly and sometimes is not seen for years, many agencies are understaffed and underfunded, and neither the pay nor the prestige is equivalent to those of other professions.

Christine Maslach, who has written extensively on burnout in the helping professions, emphasizes that burnout is a progressive process (Maslach, 1982). Most burned out workers didn't start out that way, nor did they get that way quickly or abruptly. Most interns swear they'll never be "like that." But burnout sets in slowly, usually without your knowledge. One of the major misconceptions that new workers have is that burnout only happens to "bad" workers, something they are not and will never be (Sweitzer, 1999). Burnout is relatively easy to address during the early stages, but once a person is deep into burnout, it is very hard to recover. It is easy for interns to feel judgmental about and superior to workers who are in the midst of this struggle. Of course, these workers can have a negative effect on the work at the agency in many ways, but they deserve respect and assistance. You probably cannot assist them, but you can respect them.

PATTERNS OF ADJUSTMENT

Burnout is just one possible outcome of a career in the helping professions. Many people avoid this outcome and have productive, satisfying careers. Nevertheless, over the course of many years in the field, there are adjustments to be made. There are several ways to cope with the ongoing demands—both physical and emotional—of the work and the day-to-day strains and frustrations that are as much a part of the work as the joys and satisfactions. Russo (1993) has described several patterns of adjustment that are found in experienced workers. They fall into three categories: (a) workers who identify with the clients, (b) those who identify with their coworkers, and (c) those who identify with the organization. Russo emphasizes that these are not static categories; people often move from one to another and back over the course of their careers. It is probably far too early in your own career to determine which category fits you, but knowledge of them may help you make sense of what you are seeing at your placement.

The first category, workers who identify primarily with clients, can be further divided into four subcategories. *Reformers* tend to be impatient with anything that they believe interferes with their ability to serve the clients. They often neglect paperwork and will try to change the organization's policies and procedures to better meet the needs of the client (i.e., people before paper). In our experience, this is a very common stance for interns, although they do not actively try to change the placement site. *Innovators* still promote change, but are more patient with and understanding of the change process. They listen, they ask questions, and they work with others to try to find the best way to achieve change. *Victims*, on the other hand, are frustrated by systems they see as inadequate and even harmful. They see themselves as the protectors of the clients and are likely to battle the administration openly, sometimes enlisting clients (or interns) in their struggle. Finally, *plodders* identify with the clients and may have some of the same concerns and frustrations as some of the other types described, but they seem to have given up on change. They work quietly with their clients, doing the best job they can. They often have their own way of doing things, and they work without making waves, and do not try to influence others or the organization itself to change.

Second, workers who identify primarily with their coworkers also care about clients, but in addition, they feel a strong allegiance to their profession as a whole. They may be active in unions and/or professional organizations and look to these groups, as opposed to their particular workplace, as their primary guides. Some relatively new workers are attracted to this stance because they are unsure of their own skills and knowledge. They follow the rules of the professional organization rigidly. Other workers in this category are more sure of themselves and regularly consult their union or professional organization for guidance, but consider these groups as one of several sources of wisdom.

Finally, workers who identify with the organization look to the organization and its policies and procedures as their primary source of guidance. Even though those rules may sometimes work against the needs of a particular client or be in violation of standards or ethical codes issued by professional organizations, these workers believe that following the rules will do the most good for the most people in the long run. Some may be hiding behind the rules so they do not have to think hard or take risks. Others have adopted this stance after careful thought and reflection on their experience. Russo also points out that some of them are conflict avoiders. They realize that the needs of clients, the organization, and the profession can sometimes conflict, but they want to resolve those conflicts quickly, and adherence to the rules is one way to accomplish that.

Becoming comfortable with a new group of people takes time. How much time depends on the situation you are in, the people that you work with, and the person that you are. Ideally, you will enter your relationships with coworkers with a clear set of expectations and prepared for some of the challenges of acceptance.

SUMMARY

Relationships form the primary context for your learning at the internship. As such, beginning them well is an important part of meeting the challenges of the anticipation stage. If you prepare well for them by questioning your assumptions and gathering as much factual information as possible, you will have a better chance of success. Examining your stereotypes about your supervisor and coworkers will improve your chances of a smooth entry and should reduce your anxiety as well. Acceptance is an issue in all these relationships, as it is in so many areas of life. Thinking about the issues raised in this chapter, and about those raised in Chapter 3, will let you think about and approach the acceptance process from both sides of the relationship.

Although people are of critical importance, they do not tell the whole story of an internship. To complete the picture of your internship and your journey through the anticipation stage, you must consider your placement site as an organization. That is the subject of the next chapter.

For Further Reflection

FOR PERSONAL REFLECTION:
SELECT THE QUESTIONS IN EACH CATEGORY
THAT ARE MOST MEANINGFUL TO YOU

YOUR SUPERVISOR

1. What is your supervisor's primary theoretical orientation?
2. Characterize your supervisor on the following continua: (a) instrumental versus expressive, (b) support versus direction, (c) collaborative versus hierarchical.
3. In what ways is your supervisor's style well matched to yours? Are there any areas of mismatch?
4. Supervisors have different ways of finding out how you are doing. These methods include direct supervision, self-report, and peer review. Which method or methods does your supervisor use? Which method or methods would you prefer?
5. Does your supervisor follow a structured format? If not, who decides what will be discussed? If you don't have any particular questions or concerns, will your supervisor bring up topics?
6. How are you going to be evaluated by your supervisor? What form will be used? What criteria will you be judged by? What happens to the evaluation after it is written?
7. Here is another exercise for those of you who have had other internships:
   - Compare and contrast the supervision experiences you have had. You should include, but not necessarily limit yourself to, items 1 through 6.
   - Compare your reactions to your supervision experiences. It's not always the case that one is better or worse, but they are always different, and they are bound to affect you differently.

- What are the things you need from a supervisor? If you were "shopping" for one, what would you be looking for?
- Do your positive and negative reactions to supervision fit into any patterns in your life? For example, are some of the things that make you uncomfortable in supervision also things that make you uncomfortable in other areas of life? Are you happy with these patterns?

YOUR COWORKERS

1. How would you characterize your coworkers? What is their level of skill? Experience? Professionalism? Are they similar or different from what you expected?

2. How do your coworkers seem to be responding to you? How do you feel about that response?

3. What do you hope your coworkers will be able to give you and do for you?

4. Do you recognize any of the patterns of adjustment discussed in this chapter (see pp. 127–128)?

5. Are any of your coworkers showing signs of burnout?

SPRINGBOARDS FOR DISCUSSION

1. Many interns are concerned about discussing their style and needs with their supervisors. Try using triads to role-play this conversation. Each intern should take a turn being the intern, the supervisor, and the observer. The observer's job is to give feedback to the intern on how effectively and clearly they stated their needs. The intern can then discuss what was easy and challenging, and perhaps brainstorm some other approaches and try them out.

2. Discuss, as a large group or in small groups, the risks for burnout inherent in each of your placement sites. Even though you may not be feeling burned out or anything like it, develop some strategies to keep yourself from heading in that direction.

For Further Exploration

Corey, M. S., & Corey, G. (2003). *Becoming a helper* (4th ed.). Pacific Grove, CA: Brooks/Cole.

Helpful section on supervision.

Gordon, G. R., & McBride, R. B., & Hage, H. H. (2001). *Criminal justice internships: Theory into practice* (4th ed.). Cincinnati, OH: Anderson Publishing.

Excellent discussions of issues relating to supervisors and coworkers. Obviously, this book is aimed at interns in a particular kind of setting, but it has many applications outside criminal justice.

Hersey, P., Blanchard, K., & Johnson, D. E. (2001). *Management of organizational behavior: Utilizing human resources* (8th ed.). Englewood Cliffs, NJ: Prentice Hall.

Situational leadership theory is a very popular approach to management and supervision. Discusses combinations of support and direction and when each combination may be most helpful.

Royse, D., Dhooper, S. S., & Rompf, E. L. (2003). *Field instruction: A guide for social work students* (5th ed.). Boston: Allyn & Bacon.

Excellent section on supervision.

Russo, J. R. (1993). *Serving and surviving as a human service worker* (2nd ed.). Prospect Heights, IL: Waveland Press.

Useful perspectives on supervision and coworkers. Especially thorough treatment of patterns of adaptation found in veteran workers.

Schutz, W. (1967). *Joy.* New York: Grove Press.

In his first stage of group development—inclusion—Schutz talks a great deal about acceptance concerns and the various ways of handling them.

Wilson, S. J. (1981). *Field instruction: Techniques for supervisors.* New York: Free Press.

Useful discussions of supervision and evaluation.

References

Alle-Corliss, L., & Alle-Corliss, R. (1998). *Human service agencies: An orientation to fieldwork.* Pacific Grove, CA: Brooks Cole.

Baird, B. N. (2002). *The internship, practicum and field placement handbook: A guide for the helping professions* (3rd ed.). Upper Saddle River, NJ: Prentice Hall.

Berg-Weger, M., & Birkenmaier, J. (2000). *The practical companion for social work: Integrating class and field work.* Needham Heights, MA: Allyn & Bacon.

Bogo, M. (1993). The student/field instructor relationship: The critical factor in field education. *Clinical Supervisor, 11*(2), 23–36.

Borders, L. D., & Leddick, G. R. (1987). *Handbook of counseling supervision.* Alexandria, VA: Association for Counselor Education and Supervision.

Cherniss, C. (1980). *Professional burnout in human service organizations.* New York: Praeger.

Collins, P. (1993). The interpersonal vicissitudes of mentorship: An exploratory study of the field supervisor-student relationship. *Clinical Supervisor, 11*(1), 121–136.

Corey, G., Corey, M. S., & Callanan, P. (2003). *Issues and ethics in the helping professions* (6th ed.). Pacific Grove, CA: Brooks/Cole.

Corey, M. S., & Corey, G. (2003). *Becoming a helper* (4th ed.). Pacific Grove, CA: Brooks/Cole.

Costa, L. (1994). Reducing anxiety in live supervision. *Counselor Education and Supervision, 34*(1), 30–40.

Floyd, C. E. (2002). Preparing for supervision. In L. M. Grubman (Ed.), *The field placement survival guide: What you need to know to get the most out of your social work practicum* (pp. 127–133). Harrisburg, PA: White Hat Publications.

Galassi, J. P., & Trent, P. J. (1987). A conceptual framework for evaluating supervisor effectiveness. *Counselor Education and Supervision, 26,* 260–269.

Gordon, G. R., McBride, R. B., & Hage, H. H. (2001). *Criminal justice internships: Theory into practice* (4th ed.). Cincinnati, OH: Anderson Publishing.

Hersey, P., Blanchard, K., & Johnson, D. E. (2001). *Management of organizational behavior: Utilizing human resources* (8th ed.). Englewood Cliffs, NJ: Prentice Hall.

Horejsi, C. R., & Garthwait, C. L. (2002). *The social work practicum: A guide and workbook for students* (2nd ed.). Needham Heights, MA: Allyn & Bacon.

Hunt, D., & Sullivan, E. (1974). *Between psychology and education.* Hinsdale, IL: Dryden.

Inkster, R., & Ross, R. (1998, Summer). Monitoring and supervising the internship. *NSEE Quarterly,* 10-11, 23-26.

Kanter, R. M. (1977). *Men and women of the corporation.* New York: Basic Books.

King, M. A., & Peterson, P. (1997). *Working with interns: Management's hidden resource.* Workshop presented at the annual meeting of the American Probation and Parole Association, Boston.

Kiser, P. M. (2000). *Getting the most from your human service internship: Learning from experience.* Belmont, CA: Wadsworth.

Leddick, G. R., & Dye, H. A. (1987). Effective supervision as portrayed by trainee expectations and preferences. *Counselor Education and Supervision, 27*(2), 139–154.

Loganbill, C., Hardy, E., & Delworth, U. (1982). Supervision: A conceptual model. *Counseling Psychologist, 10*(1), 3–42.

Maslach, C. (1982). *Burnout: The cost of caring.* Upper Saddle River, NJ: Prentice Hall.

McCarthy, P., DeBell, C., Kanuha, V., & McLeod, J. (1988). Myths of supervision: Identifying the gaps between theory and practice. *Counselor Education and Supervision, 28*(1), 22–28.

McClam, T., & Puckett, K. S. (1991). Pre-field human service majors' ideas about supervisors. *Human Service Education, 11*(1), 23–30.

Moses, H. A., & Hardin, J. T. (1978). A relationship approach to counselor supervision in agency settings. In J. D. Boyd (Ed.), *Counselor supervision* (pp. 437–512). Indianapolis, IN: Accelerated Development.

Royse, D., Dhooper, S. S., & Rompf, E. L. (2003). *Field instruction: A guide for social work students* (5th ed.). Boston: Allyn & Bacon.

Russo, J. R. (1993). *Serving and surviving as a human service worker* (2nd ed.). Prospect Heights, IL: Waveland Press.

Schram, B., & Mandell, B. R. (2002). *An introduction to human services* (5th ed.). Boston: Allyn & Bacon

Shulman, L. (1983). *Teaching the helping skills: A field instructor's guide.* Itasca, IL: F. E. Peacock.

Speizer, J. J. (1981). Role models, mentors and sponsors: The elusive concepts. *Signs: Journal of Women, Culture and Society, 6*(4), 692–712.

Suelzle, M., & Borzak, L. (1981). Stages of fieldwork. In L. Borzak (Ed.), *Field study: A source book for experiential learning* (pp. 136–150). Beverly Hills, CA: Sage Publications.

Sweitzer, H. F. (1999). Burnout: Avoiding the trap. In H. Harris & D. Maloney (Eds.), *Human services: Contemporary issues and trends* (2nd ed., pp. 215–230). Boston: Allyn & Bacon.

Twohey, D., & Volker, J. (1993). Listening for the voices of care and justice in counselor supervision. *Counselor Education and Supervision, 32*(3), 189–197.

Wilson, S. J. (1981). *Field instruction: Techniques for supervisors.* New York: Free Press.

CHAPTER 8

Getting to Know
the Placement Site

I learned the rules by observing my surroundings and conversations.

STUDENT REFLECTION

DON'T SKIP THIS CHAPTER!

You may be looking at the title of this chapter and wondering what is going on. After all, you have already read about and considered the clients, your supervisor, and your coworkers. You understand your role, you have a learning contract, and you have read about the stages of an internship. What else is there? Although you do know quite a bit about the placement site, you only have part of the picture. Part of your orientation and adjustment to your internship is learning how the agency or company you work for operates and why it operates that way. This knowledge will help you make sense of your experience there, although it may not seem like it right away.

In addition, if you are not prepared for the organizational dynamics and issues discussed in this chapter, you could be in for some unpleasant surprises. Suppose, for example, that you are working in a geriatric facility. Your supervisor, who meets with you once a week, is not always on the floor with you, but a shift supervisor is there most of the time that you are. This person seems unusually hostile and appears to resent the time you spend with your supervisor. She also occasionally contradicts something your supervisor has told you. While this situation would not be easy under any circumstances, it will be less mysterious if you understand that your supervisor was recently promoted to that job and the shift supervisor was passed over, even though she has been there longer. Whether you know it or not, organizational dynamics are bound to affect you, your clients, your colleagues, and your supervisor.

LENSES ON YOUR PLACEMENT SITE

We are going to encourage you to look at your placement site through the twin but related lenses of systems theory and organizational theory, each of which you may have studied in some of your classes. We have chosen a few concepts from systems and organizational theory that we think are particularly relevant to your work as an intern. We will not be covering all, or even most, of the major concepts in systems or organizational theory, and you may want to investigate them further, or even take a course in organizational behavior.

A *system* is a group of people with a common purpose who are interconnected such that no one person's actions or reactions can be fully understood without also understanding the influence on that person of everyone else in the system (Egan & Cowan, 1979). Even if you understand each person in the system as an individual, to understand the system you must understand the way everyone interacts; the whole, in this case, is greater than the sum of its parts (Berger & Federico, 1985).

Systems can be analyzed internally or externally (Berger & Federico, 1985). Internal analysis involves studying the inner workings and components of the system and the way human and material resources are arranged and expended. However, it is important to realize that all systems are hierarchical. This does not mean that they use a hierarchical authority structure; some systems do not. But each system is part of a larger system, and most can be broken down into smaller systems. A family services center, for example, can be broken down into the various programs it runs. However, it is also part of a system of service providers in the city or town in which it operates. An external analysis examines the relationships between a system and other related systems.

In both internal and external analyses, organizational theory can help you look at, make sense of, and navigate your placement site. There are many different organizational theories. For this chapter we rely heavily on the integrative work of Lee Bolman and Terrence Deal, whose book, *Reframing Organizations*, is an excellent resource for you if you want to explore organizational theory further (Bolman & Deal, 1991).

We begin with some of the components of an internal analysis. First we outline some important background information that is important to know about any organization. Then we move on to the structure of your placement, which includes the way the work is divided and various components of the agency are coordinated. Next, we help you to consider the human dimension of the organization and the ways in which the people who make up any agency go about matching their needs with the priorities of the organization. Politics are an important part of life at any human service agency and are an important and useful way to look at organizations. Finally, we will help you to look at your placement site as a culture, with symbols and rituals that can tell you a lot about organizational values.

BACKGROUND INFORMATION

In presenting the concepts in this chapter, we refer to a variety of kinds of placements, including a fictional human service agency called the Beacon Youth Shelter, an amalgamation of several agencies with which we have worked.

History

Every organization has a history, and it provides an important context for the way things are currently being done. Older organizations have often gone through a number of changes. For example, the Beacon Youth Shelter always had a rule that they would not accept residents with a history of physical violence. However, they have recently agreed to change and allow children with violent histories to come in. That means this population is relatively new to them, and there is no doubt an adjustment period going on for everyone. Can you see why this information would be useful to an intern?

Mission

It is easy to gain a general sense of the purpose of your placement; sometimes the name is all you need, as in Planned Parenthood or Child and Family Services. But each agency has an overall mission, and that is a good place to start to see what is distinctive about it. It is likely that your placement site has a written statement of its goals. Here are some examples from agencies we have worked with:

- To provide quality and multicultural services to those whose lives have been affected by sexual assault
- To provide education directed at the prevention of violence
- To serve adults and youth who exhibit or are at risk of criminal or delinquent behavior, substance abuse, or mental illness, as well as other socially disadvantaged persons

These formal mission statements may be well known to the people who work at the agency and/or to its clients, or they may be unknown to either group. A lot depends on who determined the mission and through what process (Caine, 1998). There will most likely be greater investment in and awareness of mission statements that were arrived at collaboratively, as opposed to those created by one or two people and then handed down.

Goals and Objectives

Mission statements are a good place to start, but you will find that agencies that do similar work often have similar mission statements. There is, however, a wide variety in the specific goals and objectives among organizations, even among those with very similar titles. Goals are rather broad and hard to measure. Putting an end to violence against women is an example of a goal statement. It is a goal to work toward, but not one that is likely to be reached any time soon, and therefore it is hard for the agency to measure progress toward it. That same agency might have several objectives, including a 25 percent reduction in the incidence of teen dating violence, the adoption of certain laws and procedures, and the provision of some sort of service and assistance for every woman who requests it. These are much easier to measure.

Values

Agencies also have values, or principles, that they try to adhere to in working with clients. Again, there are often formal written statements of these values, which may include such general principles as empowering clients, promoting self-sufficiency, or the right to privacy and confidentiality. Here are two examples:

- We use the word *survivor* rather than *victim* to describe someone who has been sexually abused or raped, in order to reflect the person's strength and healing capacity.

- The agency remains committed to extensive collaboration within the communities it serves. The systemic approach of the organization is best served by collaboration, given the complexity and nature of the issues confronted by our clients and in our communities.

Strategies

Each agency has a basic approach or set of approaches in accomplishing its mission. Consider, for example, a controversial program dealing with very violent children whom many other agencies and in-school programs have been unable to help. The agency's basic philosophy is that most of the time these children use violence or the threat of violence to avoid taking responsibility for themselves. If students have not done their homework, for example, and are confronted by their teachers, they may erupt. If students are acting out in class and are punished, violent outbursts may result in everyone's attention being directed toward the violence, and the original infractions are forgotten. The program also believes in giving children choices and holding them responsible.

So, if students are acting out, they are asked—once—to stop. If they do not, they are not asked again; they are given a consequence, such as being forced to stand absolutely still, in a small room, facing the corner. They can say they are sorry or that they will stop now or threaten violence, but the consequence is still given. Furthermore, a threat or a violent episode only results in longer time in the room. Sometimes these situations escalate to the point where a child needs to be physically restrained. Once the child is calm, the original consequence is instituted. You may or may not agree with this approach, but perhaps you can see how these unusual methods flow logically from the program's philosophy and beliefs. These methods, in turn, are a lot easier to understand if you are aware of and understand the philosophy behind them.

Funding

One aspect of human service agencies that is often overlooked by interns is where the money to operate actually comes from. Even if your placement site does not charge for its services, they are not free. The money to pay the staff and administer the agency that provides the "free" services has to come from somewhere. Generally speaking, agencies can be divided into three groups: public, private nonprofit, and for-profit.

PUBLIC AGENCIES

Public agencies are funded through local, state, and federal tax revenues. They may have some other sources as well, such as foundation support or grant money, but taxes provide the majority of their funds. Sometimes the clients are charged a fee on a set scale or one that varies according to their income, but those fees do not begin to cover the cost of operation. What that all means is that these agencies are ultimately accountable to the legislative bodies. That is usually how they were created, and it is typically the body that decides on the agency's budget. Berg-Weger and Birkenmaier (2000) note that public agencies tend to be large, with a vertical authority structure (we will be discussing structural issues more later on). They further note that public agencies tend to have complex rules and procedures, and a lot of accompanying paperwork. These agencies, or their parent organizations, lobby legislators to ensure that their funding is maintained or increased. Examples of public agencies include departments of child welfare, aging, corrections, and education.

PRIVATE NONPROFIT AGENCIES

Nonprofit agencies get their funding from a variety of sources, including charitable organizations such as the United Way, foundations, corporations, and individual donors. This diversity of funds is a challenge, because it means developing relationships with, and being accountable to a number of groups, but it is also a strength in that the organization may be able to avoid being overly dependent on any one source (Berg-Weger & Birkenmaier, 2000). Nonprofit agencies may receive some public funds, usually through contracts with state agencies or though Medicare. Beacon Youth Shelter, for example, contracts with the state's child welfare agency for a certain number of beds each year. Nonprofit agencies are accountable to a board of directors, who are typically volunteers. If your agency has such a board, you may find it interesting to attend a board meeting, if the agency will permit it. Nonprofit agencies generally have a lot of paperwork, but unlike public agencies it tends to go to a variety of funding sources, and sometimes to regulatory agencies as well. Fund-raising also takes up a great deal of time and energy, and as an intern you may have the opportunity to be involved in some fund-raising activities.

FOR-PROFIT AGENCIES

For-profit agencies get funds to start their work from individuals or private organizations, but they expect to be self-supporting. In fact, they expect more than that—they expect to make a profit. Kiser (2000) notes that this gives these agencies a dual mission: to serve their clients and to make a profit. Berg-Weger and Birkenmaier (2000) assert that profit is always the driving factor in these agencies. In any case, these agencies have to pay attention to making money, and they are held accountable for that goal, usually to a board of directors. Keeping client interests primary while still fulfilling their obligations to the board is one of the key challenges for these agencies. Like nonprofit agencies, for-profit agencies may access public funds indirectly, by contracting with state agencies. A for-profit agency might, for example, monitor children who are in court-mandated placements and services, and there have been some well-publicized experiments where public schools are operated by private organizations. The growth in for-profit human service organizations is also referred to as *privatization* and extends into many areas, including nursing homes, group homes for the mentally challenged, and rehabilitation.

This issue of funding is especially important because the source of funding has great power over and influence on the organization. Insurance companies, for example, usually insist on a formal diagnosis before authorizing treatment. Some human service providers are troubled by these diagnostic labels, which tend to stay in clients' files and follow them around. Nevertheless, a diagnosis must be made, and it must be one the insurance company will pay for, or the client will have to pay cash or be referred to another agency. Organizations funded by tax dollars are vulnerable to the opinions (informed or not) of the taxpayers and must worry about public relations and influencing the local political process.

Budgets

The organization's budget may not seem very important to you; after all, you are probably not getting paid! However, the financial as well as human resources that an organization has determine what it can do for clients and often affect the general tone of the workplace. Your placement site certainly has an annual budget. It may be part of a larger budget, as in the case of a senior services center that is funded by the city; its budget is part of the social services budget, which is in turn part of the city budget. The budget is usually broken down into categories, such as salaries, benefits, supplies, food, transportation, and so on. It is also important to know how the budget is set, who must approve it, and how changes are negotiated. Of course, budgets can change suddenly; in human services, that usually means they get cut, but occasionally an agency will be awarded a grant or get a new program approved and funded, which may in turn affect the agency's goals.

AGENCY STRUCTURE

An organization's structure is the way it is set up to accomplish its goals (Bolman & Deal, 1991). There are two basic elements of structure: division of responsibilities and coordination of work. There are also many subcategories within each of those two elements. Each feature of an organization's structure can be tight or loose, clearly defined or ambiguous. There are endless configurations, and no one is the best for all situations. However, some may work more effectively in certain situations than others. Let us look more closely at your placement site using a structural lens.

Division of Responsibilities and Tasks

Who will do what, and in what configurations, is one of the basic questions facing any organization. Most have developed a formal structure to answer this question, although as we will see later on, sometimes the real operation of the agency differs from the formal structure.

ROLES

Roles describe the positions in the organization and the duties or functions that each one performs on a regular basis. The formal roles in an organization can be found in two places. Most organizations have job descriptions for each position that state, at least in general terms, the responsibilities of the positions. These descriptions may be found in the policy manual. They are often generated, or reviewed, when someone is hired. A prospective employee will usually want to see it, and in a large organization, a higher-level administrator needs to see a job description to approve the hiring. Many placement sites even have written job descriptions for interns.

An organizational chart is also valuable when examining formal roles. An organizational chart does not describe the duties of each position, but it shows their position relative to one another. It shows who is responsible for whom and to whom each person is accountable. A chart for the Beacon Youth Shelter appears in Figure 8.1.

Another issue in examining roles is the degree of specialization assigned to each role. In some agencies, there is a series of highly specialized jobs and each is assigned to one person (or one category of people). For example, at a child welfare agency there may be one "role" that investigates reports of child abuse. This is an intake worker. Workers who assist victims of abuse are in a different role—they are called treatment workers. If the family needs economic assistance, they are referred to yet another worker. In other agencies there is less specialization. For example, at an agency providing services to persons with HIV/AIDS, there are several caseworkers. Those caseworkers try to handle all the needs of their clients. Of course, they cannot handle all the needs themselves and need to make arrangements with many other people and agencies, but they coordinate the services, and they know quite a bit about each one.

Both these models have advantages and disadvantages. It is a daunting task to know very many functions and sets of information well, and if someone's role is too generalized, clients may get poor advice or miss out on important options. On the other hand, when services are highly specialized, clients can get caught in a maze of workers and may feel overwhelmed and confused. It is frustrating to ask a question and be told, "That is not my area." It has probably happened to you!

GROUPINGS OR TEAMS

Roles are often organized into teams. Sometimes those groupings are functional, and people who have the same or similar tasks are placed in the same department. Other ways to organize groups are according to time (day shift versus night shift), clients (as in the case of intake versus treatment workers), and place (such as a satellite office) (Bolman & Deal, 1991).

Coordination and Control

Regardless of the particular roles and teams in an agency, there has to be a way to coordinate what they do and guide or control their efforts. How much coordination is necessary depends on the interdependence of the roles involved. In a hospital or a

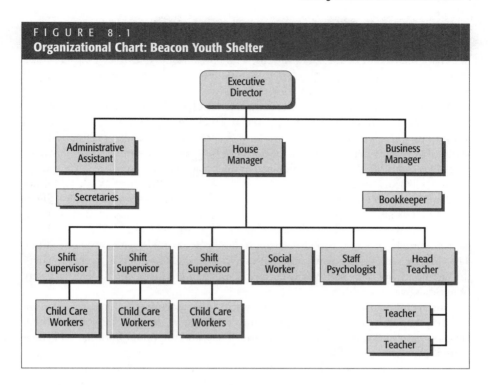

FIGURE 8.1
Organizational Chart: Beacon Youth Shelter

school, for example, the people in various roles are very dependent on one another in order to accomplish their goals, and so their functions must be carefully coordinated. According to Bolman and Deal (1991), there are two dimensions of coordination and control: vertical and lateral. Every organization has them both, but some emphasize one more than the other. And of course there is considerable variation within each of these dimensions.

VERTICAL COORDINATION AND CONTROL

Vertical coordination and control refer to the ways in which the work of the agency is assigned, controlled, and supported. This dimension consists of structures for authority (the chain of command), supervision, communication, decision making, policies, and rules.

The authority structure of an organization defines the chain of command, and this chain is visible in the organizational chart. You can see from the chart on the Beacon Youth Shelter that the top administrator is the executive director. The executive director of Beacon Youth Shelter is responsible for all phases of the operation. However, most of the actual running of the shelter is done by other people. The executive director is most closely linked to the house manager, who is in turn responsible for several professional positions. The executive director may or may not attend staff meetings or know any of the child care workers by name.

An organizational chart also gives you an easy view of the length and complexity of the chain of command. Organizations that have many layers and a long chain of command are referred to as *tall* organizations, while organizations with fewer layers and a shorter chain of command are referred to as *flat* organizations (Queralt, 1996). Each has its advantages and disadvantages.

You will notice that there is no intern on this chart. If an organization does not have an intern each semester or year, it may not include one. However, it is important to find out where you fit on the chart. Most interns tell us they fit "on the bottom." While that may or may not be true, it does not help us or the interns know to whom they are responsible or that person's place in the organization.

You will also notice that there is no place for clients on the chart. While that is not unusual, it is important to know the position occupied by clients in your agency (Queralt, 1996). In some agencies, clients are included in decision making. There may be an advisory board, for example, or a client council. This arrangement is more likely in organizations where clients can choose to use (and pay for) the service or go elsewhere. This arrangement gives clients some power and leverage. However, in agencies where the clients are not there voluntarily or have no other alternatives, clients may have very little influence on decisions.

Successful communication is, of course, the critical factor in making any organization run smoothly. In a small organization, communication may be very informal. Everyone sees and works with everyone else all day, and important information can be shared relatively easily, although it sometimes is not. Communication in larger organizations is more complicated, and often happens through memos, which is short for *memoranda*. Memos can be sent to just one person or to a whole group of people. In some organizations, employees receive several memos every day; some people call them "organizational junk mail." Increasingly, larger organizations with computer networks are using electronic mail for internal communication.

It is possible to look at patterns of communication in an agency. In some organizations, communication flows from the top levels down, with each level playing a role, and from the bottom levels up, with relatively little communication among people or departments at the same level. This arrangement is called a *chain pattern*. In a *wheel pattern*, on the other hand, one person or department is at the hub. Most communication flows directly from various departments to this hub, and from the hub to the departments. Again, there is relatively little communication among the departments themselves. In an *all channel network*, everyone communicates with everyone else (Queralt, 1996). There are many other configurations, and it can take some time to determine what pattern a given agency actually uses.

Decision making is another important structural consideration. Suppose you are working at Beacon Youth Shelter and you have an idea for a field trip. There is an interesting exhibit at the science museum, and you are successful in interesting some of the residents. But the exhibit is not going to be there much longer. Who must you ask about this trip, and how long will it take to get approval? Your supervisor thinks it is a great idea, but you are surprised to learn that the executive director, who is out of town, must approve all such trips. You also have to obtain permission from each child's parent, guardian, or case worker, and teachers must approve the absence from school for a day. This story is a bit exaggerated (although in some cases not by much), but it illustrates the importance of understanding the decision-making mechanisms at your placement and how they affect your work.

Every organization has at least some formal written rules that everyone is supposed to follow. These rules can be found in an employee handbook or policy manual, which you may already have seen. Some agencies have very thick manuals; some have more than one. You may need to ask for help in finding the most relevant material. It is also important to know how these rules and procedures were established and by whom (Caine, 1998). Sometimes rules and policies are made by a committee; other times the task is delegated to one person or written by a top administrator. The change process is also significant. Some policies have not been changed, or even reviewed, in many years. Other organizations have a regular method to review and update their policies and procedures.

It is also interesting to see how many people at the agency are really familiar with the rules. Both of our colleges have a student handbook; chances are that yours does, too. In our experience, students read that book when they have a question or when they think they may have broken a rule or been treated unfairly. Otherwise, they may never open it! At some placements, you will find that this is true for the policy manual as well.

Finally, evaluation methods are an important structural consideration. Remember that part of the vertical dimension is control, influence, and guidance. Evaluation is one mechanism that can serve those purposes. In most organizations, there is some mechanism for evaluating or assessing performance. Sometimes a new employee is given an initial evaluation after a few weeks. Periodic evaluations may be done annually, semiannually, or even monthly. Evaluations are usually conducted by the person to whom you are accountable (see the site's organizational chart).

There are many approaches to evaluation. Some methods allow the employee to set goals and then be evaluated on how successfully these goals have been reached. Other times the criteria and goals are set by the supervisor, and there are lots of approaches between these two extremes. Some organizations use self-evaluation or peer evaluation in addition to or in conjunction with the supervisor's evaluation. The format of the report can also vary from a simple checklist or set of numerical ratings to more complex narrative formats.

The way in which staff members are evaluated says a lot about what is important to the organization. One important issue here is the aspects of the job that are included in the evaluation. Think about your classes in college. You have readings assigned, and many of your instructors may tell you that class participation is important. However, if you are never tested on your understanding of the reading and participation is not a part of your grade, you may be tempted to let these areas go, especially if they are difficult. By evaluating you on certain components of your performance but not on others, the instructor is sending a message, perhaps unintentionally, about what is important in the class.

Returning to your placement site, staff members are supposed to be evaluated based on how well they perform the duties in their job descriptions. In practice, however, some of these duties may not be reviewed, commented upon, or counted toward promotion, pay increases, and other organizational rewards. In addition, a supervisor may comment on aspects of the job that are not in the job description. One worker we know was criticized in writing for not being sociable with others at the agency, even though his cases were handled quite well.

Some organizations evaluate performance, some evaluate outcomes, and some evaluate both (Caine, 1998). Again, an analogy to school may help illustrate this distinction. Teachers of a U.S. history class may be evaluated based on whether they cover all the material by the end of the year and whether students seem to enjoy and be stimulated by the class. Another approach would be to evaluate teachers based on what students are learning. Some schools believe that it is the teachers' responsibility to lay out the information and the students' responsibility to learn. Others believe that the school and its teachers have a responsibility to see to it that every child learns. The evaluation methods chosen reflect these two approaches. Similarly, a human service agency may evaluate employees based on the number of clients seen or the number of hours worked, or it may choose to find ways to measure the change in clients and evaluate its workers in that manner.

LATERAL COORDINATION AND CONTROL

While every organization has a vertical structure, those are not always the most effective means of getting the work done. Policies and chains of command are not always followed, nor are they always efficient. Lateral methods of coordination and control are those structures that emerge which are less formal and more flexible than vertical ones (Bolman & Deal, 1991). Don't confuse lateral control with a "flat" authority structure. Flat structures are less hierarchical, but lateral methods are still needed. Lateral methods may include meetings, which can occur monthly, weekly, or daily. Some agencies begin each shift with a meeting. Various staff groups then meet once a week for more in-depth discussion, and the whole group meets once a month.

Task forces or temporary work groups are another form of lateral coordination, and ones that you may have the opportunity to be involved in as an intern. Most human service agencies have to be evaluated by local, state, or even federal licensing organizations from time to time. Generally, the organization has to gather data and write about itself for the visiting team; this is called a self-study. In addition, the team visit, which can last for several days, needs to be organized and coordinated, and large numbers of documents have to be made available. Some of this work can be done on an ongoing basis and built into someone's role. But some of it is temporary, and occurs only once every few years. In these cases a temporary team may be organized to prepare the self-study, arrange documents, and handle the logistics of the team visit.

Task forces usually arise out of some particular problem or challenge. For example, the local Boys and Girls Club may be concerned about the lack of racial and ethnic diversity in their clients, feeling that it does not represent the diversity in the surrounding community. A task force, consisting of individuals from many segments of the organization, may be appointed and asked to study this problem and propose a solution.

HUMAN RESOURCES

Another way to look at your placement site is to focus on the people rather than the structure. In doing so, you will see a whole other, and very rich, dimension of life in that organization (Bolman & Deal, 1991). Here we are concerned not so much with what people are supposed to do, but with what they *actually* do as they try to match their own needs to the needs and features of the organization. As with the structural lens, there are a number of components to consider when assessing the human dimension.

Communication Skills

Using the structural lens, you looked at patterns of communication. These patterns emerge over time, and it is almost like you must observe from a distance to see them. Now we turn to the skills that individuals and groups show in communicating with one another. You have more than likely studied communication skills in your academic program, so this will be a review.

One aspect of communication is its *clarity*. Consider how well people in the organization listen to one another and how they are helped to do that. Another aspect of communication is its *openness*. In some organizations, people generally say what they think, about issues and about one another, in an honest and clear way. There is no prescription for how to accomplish that goal; specific approaches vary depending on the constellation of individuals and cultural profiles in the organization. Still, every organization needs to find a way to accomplish this goal, and some do so far more effectively than others.

Conflict resolution is another set of communication skills that is visible in various ways and to various degrees in organizations. If there are not some structured, accepted, and supported ways of dealing with conflict in place, more informal and often less effective methods will be used. One of us had an intern whose coworkers complained about her in formal and informal meetings, yet no one spoke directly to her, nor did anyone try to stop the conversations that were taking place behind her back. When someone finally took her aside and told her of these conversations, she spoke with her supervisor. The supervisor then went to the individuals involved, but not to the group, and not in the presence of the intern. Needless to say, the situation did not get better.

Norms

As you read the section on the rules of organizations, you may have occasionally wondered how to answer some of the questions about your placement. In some cases, you don't know the answer at all, but in others, you may be wondering, "Do you want to know what's written down or do you want to know what's really going on?" Often, the formal rules, roles, goals, and values only tell part of the story. Consider the speed limit on the interstate. The formal limit may be 55 or 65 mph, but how fast you can go before risking a ticket is the informal rule that most people concern themselves with. If you live in a college dormitory, there is probably a time after which you are supposed to be quiet. But the unwritten rule usually is that you can make all the noise you want as long as no one complains. These are examples of informal rules or norms that contradict or modify the formal rules. Other norms are not products of the formal rules. Most schoolchildren can tell you that it's wrong to "rat" or "tattle," often to the chagrin of adults. This is a very powerful rule for many children, yet it is not written anywhere.

Every organization has unwritten rules. Think back on jobs you have had and on your current placement. There is an official start time, but there is usually a grace period of at least a few minutes. In some places, you are expected to work late—for no extra pay—and in other places, you are not supposed to do that even if you want to. Many organizations have an informal dress code to go along with the formal one. Sometimes the informal rules come about because the formal rules have eroded or became obsolete, but have never been changed (Gordon, McBride, & Hage, 2001).

At one family services agency, for example, the hours were 9 A.M. to 7:30 P.M., Monday through Friday. However, in recent years, the agency has been doing more and more outreach programming, working with groups in the community, meeting during the evening in church halls or people's homes, and developing a lot of weekend programs. Fridays are an especially busy day, as the finishing touches are put on the weekend activities. As a result, Mondays have become an unofficial day of "downtime." Few appointments or meetings are scheduled, and many staff members do not come in until noon. If you were an intern there and wanted to work on Monday or tried to launch a new group that meets on Mondays, you would receive a chilly reception, and you might wonder why. Here are some other examples of unwritten rules that interns have encountered:

- Go to lunch with your coworkers or they will think you are a snob.
- Never go out to lunch; work at your desk.
- Don't miss the director's holiday open house.
- If you are dating another staff member, keep it quiet.

The written rules are usually easy to find, but how do you go about finding the unwritten ones? First of all, you will usually know when you have broken one. Just like in school, when you wore the "wrong" thing, used an expression that was suddenly "out," or sat at the "wrong" table in the cafeteria, people will react strangely; perhaps someone will even take you aside and clue you in when you violate an unwritten rule. Another way to discover the unwritten rules is to know the written rules and observe what happens when they are broken (Caine, 1998). In some cases, nothing happens. Either the rule has been bent in a way you don't know about, or it has been ignored altogether.

Informal Roles

There is also a network of informal roles and relationships. It will not take you long to notice that some people perform duties that are not in their job descriptions. It may be that they were asked to do it and they didn't feel that they could say "No." Consider, for example, the worker whose supervisor discovers that her route home takes her past where his son has piano lessons. One day, the supervisor asks if she would mind picking up his son, since he will be in a meeting. The worker is happy to help, but it soon becomes an expectation.

Another way job descriptions expand is when something is not in anyone's job description; someone just decides to do it, and after a while everyone relies on this individual. There may be one person, for instance, who remembers everyone's birthday, takes up a collection for a gift when someone gets married or retires, or makes the arrangements for parties. Other roles in organizations might include the jester, who makes everyone laugh, and the go-between, who seems to have access to just about everyone.

Sometimes people expand their job description to make their jobs more interesting, especially if they have had the same job for a while (Gordon, McBride, & Hage, 2001). At the Beacon Youth Shelter, a child care worker became interested in grant writing, volunteered to work on some grants, and eventually became quite good at it. Of course, his job description expanded and his paycheck did not, but it did make him a more influential person in the organization, which is another reason people sometimes try to expand their roles.

Cliques

Finally, there are informal subgroups or cliques in many organizations. At Beacon Youth Shelter, for example, there is a group of child care workers who have been there a long time; they are referred to as "the veterans." Few of them had any formal education to prepare them for their work; they learned on the job. Some of them rose in the organization to become shift supervisors, and until recently, the house manager position was held by a former member of this group. In recent years, though, there have been more and more workers hired who were graduates of programs in human services or social work. They are often younger and full of new ideas. Some of these ideas work, but others don't. The veterans sometimes shake their heads at these "whiz kids," and sometimes the whiz kids get impatient with the veterans.

Of course, both groups have a valuable perspective to offer the organization, and most of the time they work well together. Still there is some tension and competitiveness between them. It doesn't help that the new house manager did not come from within, is a graduate of a human services program, and has a master's degree. The veterans miss the easy relationship and open access they had with the old house manager; they could just approach her privately and accomplish a lot. Not only do the veterans not have this relationship with the new manager, but some of the whiz kids seem to be developing such informal access.

Management Style

Just as your supervisor has a style with you, people in supervisory positions have their own styles with the people or groups they supervise. There is considerable variation among individuals, of course, but there is often a prevailing style in the organization itself. If you go back and re-read the section on supervisory styles, you can think about how this applies to your agency.

Staff Development

Most organizations put some time and energy into helping their staff grow as professionals and do their jobs better. These activities are often referred as *staff development*. Staff development may take place during a portion of regular staff meetings, or special meetings may be held exclusively for that purpose. Some agencies take their staff away on retreats. Others send them to workshops offered by colleges or training organizations.

Still others support their staff in continuing their education. The connection to values is not just in how much time and money is put into staff development, but in the purpose of the activities (Caine, 1998). Some activities are directed at helping people learn to do their current jobs better. Others are directed at learning a new set of regulations, policies, or procedures. Still others are aimed at helping the staff learn more about theories so that they can come up with and share applications to the organization's work.

Finally, some staff development activities are directed at the career development of the staff. These activities are undertaken with the understanding that most of the staff will not and should not stay in their current jobs. They may move up in the agency or they may move on to a parallel, or better, position in another agency. Staff development of this kind is an investment in an individual and the profession as well as in the current functioning of the agency.

ORGANIZATIONAL POLITICS

Your age, your interests, and your work experience affect the associations you have with the word *politics*. In our experience, many interns associate that word only with elections. Others apply the term more broadly but find that the word has negative connotations of wheeling, dealing, and back-biting. In fact, we have had a number of students tell us that they want to go into the helping professions to "get away from all the politics" that they imagine exist in the corporate world.

In reality, you cannot escape politics. They are everywhere, and they affect you whether you see them or not. The politics of an agency can be difficult for an intern to see, especially if you are not there for very many hours each week or for very many weeks. Still, the more you can learn how to look at your placement, and organizations in general, in this way, the more successful you are likely to be.

Bolman and Deal (1991) call this lens the "political frame." When you look through this lens you see organizations as coalitions, made up of different individuals and groups. They may be formal groups or the informal groups and cliques described earlier. Those individuals and groups have substantial differences in perspective, beliefs, and priorities. That diversity can and should lead to a richness in the decisions made in the organization, but it is a mistake to think that the differences can be resolved or integrated all the time.

According to the political frame, when resources are scarce, decisions about them are critical. In most human service agencies resources are always scarce. Scrambling to raise money and to adjust to reduced allocations from legislatures, shortfalls in fundraising goals, and escalating costs are just facts of life in human services. Therefore, conflict, bargaining and negotiation are inevitable and not necessarily bad. So, the first thing to look for when you examine your placement site from this perspective is the various groups and the resource issues over which they compete. The next thing to look at is how each group accesses and uses power and influence.

Formal authority and power are not the same thing. Sometimes in organizations individuals or groups who are "lower" on the organizational chart have more power and influence than some of those above them. Secretaries often appear at the lower levels of the chart; in theory, they have little power and influence. However, it is often secretaries who know where everything is and how to navigate the paperwork maze. They overhear a lot of conversations, and they often control important resources, such as supplies. Often, if you want to know how to get something done, you are better off consulting an experienced secretary than the policies and procedures manual.

There are a number of ways to have power and influence in an organization (Bolman & Deal, 1991; Homan, 1999). Position power is one of them, but power also comes from information and expertise. If your agency is using a new set of software programs or a new computer system and you have someone who is very knowledgeable about them, that person has enormous power and influence. Pity the person who has insulted, slighted, or ignored the tech expert; her computer will have a problem sooner or later, and it will get fixed later, not sooner. Control of rewards, resources, and "inside" information is another source of power, as we noted in the case of secretaries. People who are well connected to those who control these aspects of agency life generally have what they need and know what they need to know, often in advance of everyone else. Finally, individuals can have personal power. This power may come from high levels of charisma, from natural charm and sociability, or by carefully building networks of friends and attending to each of them.

Power in organizations can also be found in the informal networks of influence. In some instances, people on the staff are related (or were related) to each other. In other instances, they have shared decades of connection through schooling, neighborhoods, or churches. In still others, outside activities such as softball may unite them. Determining these informal influence networks takes time and reflection. You might try arranging everyone in order according to the influence they seem to have on decisions or on daily operations. Russo (1993) suggests that you interview people in the organization and ask them to whom they give advice, to whom they give orders, and from whom they accept advice and orders. This may be impossible in practice, and not necessarily well received by the staff, but if you could do it, you could then construct a web of connections that would tell you a lot about how power and influence flow in the organization.

ORGANIZATIONS AS CULTURES

A final way to look at your placement site is as a culture, which, like any culture, is understood at least in part by its symbols and rituals (Bolman & Deal, 1991). When you look at behavior in organizations this way, you are looking not so much at the function of certain practices, but at their *meaning*. Commencement exercises are really not very functional if you think about it. Diplomas could be mailed to graduates, and it would be a lot easier and cheaper. But the ceremony means a great deal to the graduates and to the faculty, staff, and parents. It is a celebration of the values of the institution, a way to say, "Look, here is what we do well." It is an important annual reminder, in the face of the day-to-day frustrations and aggravations, of why teachers, administrators, counselors, coaches, and others in schools do what they do. Ceremonies marking thirty days—or five years—of sobriety in Alcoholics Anonymous are another example.

Less dramatic events, such as staff meetings, can be looked at in this way as well. They may not be the most efficient way of communicating critical information, or even the most productive forum to discuss difficult issues, but they are a way for people in the organization to see each other visibly interacting and inhabiting the same space.

Like other organizations, your placement site may have smaller, less formal rituals as well. Going for pizza on Thursday, or stopping for coffee on the way back from a site visit, or gathering in the lounge during a break may seem silly to you, but you ignore them or decline an invitation to participate at your peril. These rituals symbolize collegiality, concern, and cohesion.

Language is another aspect to the symbolic or cultural frame, and another key to operational values. For example, consider the way staff members talk about clients and the labels they apply to them. Sometimes staff members will refer to their clients in openly derogatory terms, like "retard," "dummy," or "geezer." Other times the labeling is more subtle. For example, sometimes certain clients are referred to as "difficult" or "hard cases." When we hear this, we are always tempted to ask: Difficult for whom? This comment is made using the experience of the staff as a base for description, not the experience of the client. The same client could be described as "having a hard time accepting authority" or "unwilling to accept personal responsibility." These terms are much more descriptive of the *client's* experience. Of course, people are going to become frustrated and say negative things from time to time, although preferably not to the client. A pattern, however, of one sort of comment or label can be a clue to an unconscious organizational value.

The internal workings of any organization are a complex and fascinating thing. It will take you some time and energy to learn to look at organizations in the ways discussed here. What we hope for is that you learn to do it better than you do now, and that you come to see the value of taking time for these analyses. Still, the picture is not complete. Just as it is impossible to understand a system without understanding all its component parts and relationships, it is impossible to understand that system without understanding its relationship to other systems and to larger societal forces and pressures.

THE EXTERNAL ENVIRONMENT

You may already be seeing how the world outside the agency has an impact on its functioning. The number and nature of the organization's relationships with other organizations and agencies, sometimes called the *task environment* (Queralt, 1996), will affect your work and the ability of your placement site to do its job. In a more indirect but no less important way, the surrounding community, the economic climate, and some political issues affect you, your clients, and your agency (Queralt, 1996).

The Task Environment

Most organizations cannot function without smooth relationships with other organizations. This is certainly true in human services. For example, the Beacon Youth Shelter works with two state agencies that refer and pay for residents. A local food bank helps supply their food. A large clinic in the area contracts with Beacon for certain psychological services. Some of the students attend school off grounds. Occasionally, a resident runs away or causes some problem in the community; good relationships with neighbors and the local police are essential. In addition, residents do not stay at Beacon forever; sooner or later, a new placement must be found for those who cannot return to their families or live on their own. This aftercare placement process involves detailed knowledge of and good relationships with a variety of other placement sites.

Of course, other agencies can be sources of competition as well as collaboration. That is particularly true in the world of for-profit agencies, but it is not limited to them. If you are interning with a nonprofit, other agencies in your area that provide similar services may compete for grants, state contracts, corporate sponsors, and so on.

Furthermore, most human service organizations have one or more agencies that monitor them in some way. Some agencies are part of a larger parent organization. That is not the case with Beacon Youth Shelter, but other shelters in the city are funded by organizations like Catholic Charities or the Salvation Army.

The Sociopolitical Environment

There are a number of political issues that affect placement sites; we will just mention a few of them that seem to be common across many different placements. Local politics can have a considerable impact. For example, in some states, there is a cap on property taxes, which make up a large portion of each town's budget. Towns can vote to exceed the limit, and these votes are often taken to provide a special service, such as a new gymnasium or road repairs. If your placement is heavily funded by the town, taxes and local politics are important issues for your agency.

The attitudes of the people in the town toward your agency, its clients, and the work you do are important. Some agencies, like group homes, seem to inspire at least apprehension, if not fear, among neighbors. People agree on the need for them, but want them located somewhere else. If your agency has just opened in a hostile or unwelcoming community, it has a large public relations task ahead of it. Some agencies make deliberate efforts to involve people in the community in their work. This is especially important when the clients are residents of that community, as they are in a Planned Parenthood Center, for example. If these residents are suspicious of and hostile toward the agency, chances are that fewer clients will come for services.

State and federal politics can have an impact as well. A new governor or mayor often means new people in charge of government agencies, and sometimes means a change in philosophy, along with new regulations and changes in the way funds are distributed. Federal issues such as welfare reform and abortion rights may also have a large impact on your clients and the mission and goals of your placement site.

SUMMARY

In this chapter, we have introduced you to some ways of thinking about organizations. Although you may feel somewhat overwhelmed by the information, we have really only scratched the surface. Systems theory, organizational dynamics, organizational development, values, and staff development are enormous topics. In our experience, many interns find these topics fascinating. Others do not find them so interesting. However, we believe that the concepts we have introduced here, at a minimum, will help you understand both your placement site and what is happening to you there. At the very least, this chapter should stimulate some thought now and be a resource if you begin to have problems at the placement.

For Further Reflection

FOR PERSONAL REFLECTION

As you can see, there are many aspects to an organization. The questions below are broken into categories. Some of them will require that you ask questions and do some research. Choose the questions within each category that seem most meaningful to you:

BACKGROUND INFORMATION

1. How many years has your agency existed? Have there been any major changes during that time? Describe them.
2. What is the agency's overall goal and philosophy concerning the clients? Working with clients? Managing employees and interns?
3. What is your agency's annual budget? What categories is it broken into?
4. Where does the money come from? Do clients have to pay? If not, does some other agency pay? If the services are "free," where does the agency get the money for payroll and operating expenses?

STRUCTURE

5. Obtain or create an organizational chart of the agency. Are interns on the chart? If not, place them in appropriate locations.
6. Where are the formal rules of the agency located? Have you looked at this resource?
7. How are decisions supposed to be made at your site? What evidence do you have that decisions are or are not actually made that way?

HUMAN RESOURCES

8. Have you observed any informal rules? Do any of them conflict with the formal rules?
9. Have you observed instances of people performing tasks and functions that are not in their job descriptions? What do you know about how this happened? How about informal roles such as "office clown," which aren't in anyone's job description? Do there seem to be informal or unwritten expectations of you as an intern? How do you feel about those expectations?
10. Are you aware of any cliques or informal subgroups that exist at your agency?

POLITICS

11. What are the various subgroups within your agency that may be vying for power or resources?

12. Which of the sources of power identified in this chapter can you see being used in your agency?

ORGANIZATIONAL CULTURE

13. What are some important symbols and rituals at your agency?

THE EXTERNAL ENVIRONMENT

14. What are some of the agencies with which your site has important relationships? What do they do?

15. Are there any agencies that monitor or control your agency? Are there periodic site visits?

16. What have you been able to learn about the relationship between your agency and the surrounding community?

17. What local, state, or national political issues are affecting your agency and its clients?

SPRINGBOARD FOR DISCUSSION

Analyzing organizations is not easy, and some of the concepts in this chapter may be abstract or difficult to understand. If so, one thing you could do as a class is to analyze the school you go to, using these various frames. It will be fun and it will help you learn.

For Further Exploration

Berger, R. L., & Federico, R. C. (1985). *Human behavior: A perspective for the helping professions* (2nd ed.). New York: Longman.

A thorough discussion of systems theory, including many concepts not discussed in this chapter.

Egan, G. (1984). People in systems: A comprehensive model for psychosocial education and training. In D. Larson (Ed.), *Teaching psychological skills: Models for giving psychology away* (pp. 21–43). Pacific Grove, CA: Brooks/Cole.

Egan, G., & Cowan, M. A. (1979). *People in systems: A comprehensive model for psychosocial education and training.* Pacific Grove, CA: Brooks/Cole.

A wonderful book that integrates a systems and individual development approach to human services. Excellent chapters on systems. Unfortunately, the book is out of print. If your library does not have a copy, try the preceding Egan reference.

Gordon, G. R., McBride, R. B., & Hage, H. H. (2001). *Criminal justice internships: Theory into practice* (4th ed.). Cincinnati, OH: Anderson Publishing.

An excellent chapter on organizations from a criminal justice perspective. Covers both public and private settings.

Queralt, M. (1996). *The social environment and human behavior: A diversity perspective*. Needham Heights, MA: Allyn & Bacon.

A social work text with an excellent chapter on organizations.

Russo, J. R. (1993). *Serving and surviving as a human service worker* (2nd ed.).
Prospect Heights, IL: Waveland Press.

Interesting chapter on organizations, with examples from criminal justice and mental health settings.

Senge, P. M. (1990). *The fifth discipline: The art and practice of the learning organization*. New York: Doubleday/Currency.

Very interesting and accessible discussion of systems theory and organizations. The applications are business oriented, but the concepts are very clearly explained.

References

Berger, R. L., & Federico, R. C. (1985). *Human behavior: A perspective for the helping professions* (2nd ed.). New York: Longman.

Berg-Weger, M., & Birkenmaier, J. (2000). *The practical companion for social work: Integrating class and field work*. Needham Heights, MA: Allyn & Bacon.

Bolman, L. G., & Deal, T. E. (1991). *Reframing organizations: Artistry, choice and leadership*. San Francisco: Jossey-Bass.

Caine, B. T. (1998). Using Bolman and Deal's four frames as diagnostic tools: Key concepts and sample questions. In B. T. Caine (Ed.), *Understanding organizations: A ClassPak workbook*. Nashville, TN: Vanderbilt University Copy Center.

Egan, G., & Cowan, M. A. (1979). *People in systems: A comprehensive model for psychosocial education and training*. Pacific Grove, CA: Brooks/Cole.

Gordon, G. R., McBride, R. B., & Hage, H. H. (2001). *Criminal justice internships: Theory into practice* (4th ed.). Cincinnati, OH: Anderson Publishing.

Homan, M. S. (1999). *Promoting community change: Making it happen in the real world* (2nd ed.) Belmont, CA: Wadsworth.

Kiser, P. M. (2000). *Getting the most from your human service internship: Learning from experience*. Belmont, CA: Wadsworth.

Queralt, M. (1996). *The social environment and human behavior: A diversity perspective*. Needham Heights, MA: Allyn & Bacon.

Russo, J. R. (1993). *Serving and surviving as a human service worker* (2nd ed.). Prospect Heights, IL: Waveland Press.

CHAPTER *9*

Getting to Know the Community

INTRODUCTION

In the previous chapter we encouraged you to widen the circle of your thinking, to stretch beyond your clients and colleagues and consider the placement site itself. In this chapter we encourage you to widen the circle of your thinking even further, and consider the community context of your work. In our experience, we've found that some interns are naturally drawn to the community context, others are led to it by their experience with their clients and clients' struggles, and some never really feel drawn to learning about the community. However, all interns and all internships are affected by the community and its dynamics.

In this chapter we spend some time helping you understand why the community context is important. We also spend a little time on definitions and categories of communities. Finally, we help you consider some of the important aspects of any community, and how you might go about learning this information in your particular situation.

WHY SHOULD I STUDY THE COMMUNITY?

First of all, for some of you, the community *is* your client. If your agency helps to organize community discussions of neighborhood or municipal issues, or if you are helping a group of people mobilize for political advocacy, you are working with individuals and groups, of course, but the main target of your work is the community itself. If that describes your work you probably don't need much convincing about studying the community; for you, this is as fundamental to your work as thinking about individual clients is to other interns. Indeed, you may have found yourself a bit impatient as you read Chapter 6, on clients, since it focused almost exclusively on individuals.

However, even if your work does not focus on the community directly, it is important to take some time to think about the community. It is all part of learning to think in systemic terms, something you have probably studied in your program, and something that we encouraged you do to in considering your family and the agency. For many people it is a hard lens to acquire, but it just takes practice and persistence. Once you do acquire it, though, it is equally hard to take it off! When you learn to see the world in this way, things will never look the same.

For one thing, your clients live their lives and struggle with their issues in a larger context. Communities shape the contexts of people's lives in many ways. You should not ignore some of the larger reasons that your clients have the problems that they do (Homan, 1999a; Rogers, Collins, Barlow, & Grinnell, 2000), nor the larger arenas in which assistance or solutions may lie. You are probably accustomed to thinking at least a little bit about your clients' families, even if they are not the main focus of your work, but the clients' context goes beyond that. One of our students, for example, worked with adults who were developmentally disabled. After a while, she became quite frustrated at how few of them lived independently and had jobs. She believed that they would feel much better about themselves and probably function at a higher level that way. She learned that attempts to build group homes or get employers to accept adults with developmental disabilities were often delayed or scuttled by the real or perceived attitude of the community at large. She also learned that her agency had done very little to try and educate or influence that community. Some of her clients were frustrated too, and she wondered whether it was right to spend her energy trying to help them feel better about and adjust to something she believed was unfair.

For another thing, your agency does not exist in a vacuum any more than your clients do (Horejsi & Garthwait, 2002). Many agencies grew directly out of community needs. You may remember Steven and the shelter he worked in, which we described in Chapter 6. That shelter was created in direct response to the need in the homeless community, and the community at large, to service homeless persons with substance abuse problems. The range of agencies and programs that have grown to meet the needs of persons with HIV/AIDS is another example. Regardless of their origins, though, all agencies are affected by the rhythms of the community. For example, even though Steven's shelter meets a need of the community, there are some people in the community and in the neighborhood who are adamantly opposed to it. Some of them are philosophically opposed to the approach that the agency uses; more are concerned with the safety and general appearance of the surrounding neighborhood. People concerned with attracting visitors worry about the effect of a homeless population. When those factions gain power and influence, or when a high profile crime or death occurs, efforts are renewed to close the shelter where Steven works. Other communities make similar efforts to close programs for the homeless and even conduct sweeps of people living on the streets, in the hope that the homeless will simply go away.

Finally, communities are powerful. They can be powerful allies or powerful opponents. One of us once worked with a local school to create a series of parent workshops. The topics were chosen based on input from school counselors and social workers, who in turn were listening to the parents. Transportation was arranged and child care was provided. Flyers were sent home to parents. Almost no one came. What we had not done was to find the people who were influential in the parent network in the community and enlist their help, or better yet, their input from the beginning. They did not oppose our work, but they did not support it, either. We later learned that our university, like many others, had a long history of offering programs to the community, and a long history of not delivering what was promised or pursuing its own as opposed to the community's agenda. The parents did not know us, and their general instinct was not to trust us.

Communities are important factors in your work for all these and many other reasons. So let us help you identify a community to study and suggest some of the things you may want to find out.

WHAT IS A COMMUNITY?

The first step in this process is to identify the community, and this is sometimes not easy. For one thing, we use the word *community* to refer to different kinds of groups (Homan, 1999a; Horejsi & Garthwait, 2002). Sometimes we are referring to a geographical area—a city, a town, a neighborhood, or even a block. Other times we are referring to an age group, as when we refer to the community of senior citizens; or an interest group, such as local environmentalists; or even a cultural group, such as the Caribbean community. Members of these groups may live in different parts of the geographical community. To complicate matters even further, most communities can be further divided into smaller communities.

For another thing, the community where your agency is located may be a different community than the one in which your clients live. In that case, it is probably important to know about both communities. In other cases, the clients that come to your agency come from a wide range of geographical locations and are members of many different communities of interest, age, or culture. So, if you work in a clinic that provides services to persons with HIV/AIDS, the community where the clinic is located, the communities where your clients live, and the community of people with HIV/AIDS are all communities that have impact on your work and your clients.

We say all this not to confuse you or to undertake an abstract theoretical exercise. The point is to get you to think about all the communities there are for you to consider and to acknowledge the difficulty in choosing one or two. Chances are, though, that one or two is all you will have time for. We encourage you to consult your instructor and site supervisor in identifying the community or communities that are most important in helping you and your agency achieve your goals.

COMMUNITY INVENTORY

Communities are complex and fascinating, and it takes time, experience, and some expertise to get to know them. We cannot begin to cover all of the important aspects of communities or working in them in this short chapter, but we will suggest some aspects of communities for you to consider and explore. We tried to choose facets of the community that seem especially relevant to you as an intern. The ideas here are drawn from our experience and from the work of Mark Homan, author of two seminal texts on community development (Homan, 1999a, 1999b), as well as Horejsi and Garthwait (2002), Rogers et al. (2000), and Kiser (2000). We begin with some basic information that you need to know, and then move on to ideas about analyzing communities. We have organized this analysis into the same four categories that we used in discussing organizations, drawn from the work of Bolman and Deal (1991): structural; human resources; symbolic; and political. First, though, a few words about assets and needs.

Assets and Needs

When considering any community, it is easy to think in terms of needs. And to some extent that makes sense; all communities have needs, and it is important to look at how well and in what ways the needs are met. It is easy, though, to move from "needs" to "needy," to look only at what the community does not have or does not do well. That is especially true if your interest in the community is stimulated by a desire to understand problems being experienced by individual clients. Working with youth in a Boys and Girls Club in the inner city, for example, it is easy to look at crime rates, poverty, and poor transportation as factors contributing to the problems your clients have.

When you think of communities this way, you are equating *needs* with *deficits*. It is all too easy, then, to miss the strengths or assets of the community. And that is especially easy if the community meets its needs in ways that are unfamiliar to you. Some writers in the field of community development are trying to reframe discussions of communities to focus on *assets* or *capital* (Homan, in press; Kretzmann & McKnight, 1997). As we proceed with our discussion of community inventories, we are going to try to use the language of strengths and assets, and we encourage you to do the same.

Basic Information About Communities

If you have chosen a geographic community to study, begin by learning its dimensions. How large is it, and what are its recognized boundaries? A boundary can be a town line, a street, a river, or anything that people generally accept as marking the edge of a community. Next, consider the general appearance of the community. What sort of impression does it give at first glance? What catches your eye? Are there visible local landmarks? How about landmarks you might miss at first, but that every resident knows about (like a controversial cell phone tower)?

Next, consider some basic information about the people in the community. How many people live there? Is the population dense or spread out? Is it stable or does it fluctuate? Communities with seasonal workers, summer residents, or colleges are examples of communities where the population changes over time. Finally, what is the average income of the people who live there? How does it compare to state or national averages, and to the cost of living in the community?

Structural Considerations

This category of assets concerns the formal and informal structures that allow people in the community to meet their needs, and how well they work. We begin with basic human needs such as food, shelter, clothing, and medical care. And the issue here is *access* to these assets across a variety of incomes and circumstances. What provisions has the community made to assure that these needs are met? How successful are they? For example, how does the community handle those who are temporarily or permanently displaced from their homes? Some people will have insurance for such circumstances, but some will not. What then? Are there shelters? Perhaps the community relies on networks of family and friends to handle situations like this. Does that work? Another example is grocery and other retail stores. How easy are they to find and to get to?

Another set of assets has to do with education. What is the quality of the schools in the community? What about adult education and training opportunities? Is accessible and affordable post-secondary education available? And don't forget about the informal learning, mentoring, and educational networks that exist in many communities.

Employment is another basic need, and opportunities for employment are a community asset. Some communities have a variety of employment opportunities available, while others have almost none. Sometimes people in the community have to travel a long way to get work. Also, sometimes the jobs that are available require special skills, and there is no training available. As you learn about these features of the community you are studying, the unemployment statistics (which you should also determine) will have more meaning.

Communication is also a basic need; people need to know what is going on in their community and in the world around them. Most communities have both formal and informal means of communicating necessary information, and these are powerful community assets. Identify the major newspapers, TV and radio stations available in the community. Do those media outlets provide adequate coverage of news and events in the community you are studying? Large metropolitan newspapers often overlook issues in outlying areas or in certain segments of the city. Some communities have their own means of communication. Some have newspapers or newsletters of their own, and many have gathering places where people go to catch up on what is going on. Finding these local resources is crucial.

There are also basic systems in communities that meet needs or provide access to the services described above. Adequate public transportation to, from, and within the community makes an enormous difference in people's abilities to meet their needs, as do the quality of roads, sidewalks, crosswalks, and walk lights. In some communities, clean air and water are taken for granted. Others struggle with pollution or even with a lack of potable water. Similarly, you probably don't think very much or very often about waste removal and drainage systems unless you have been in places where sewers back up, garbage overflows, and water sits in stagnant pools. Not only do these problems detract from the general attractiveness and desirability of a community, they each pose serious health risks.

Human Resources

People, of course, are an enormous asset, and the idea here is to get to know something about the people in the community, their social and emotional needs, and their strengths as well as their liabilities. In some communities it is easy to see that people feel safe and secure. They move about freely and easily; they congregate in a variety of places. In other communities people are afraid to go outside or to gather in certain places. And of course, there is lots of room between these two extremes. Pride is another social and emotional issue. Find out what people who live in the community are proud of about where they live or about the members of their community.

In Chapter 8, on organizational issues of the placement site, we mentioned operational values and told you that you can learn a lot about an organization by how it spends its time and its money. The same is true for people in the community you are studying. When people have free time, where do they go and what do they do? There will, of course, be a range of answers to that question in any community, but you can probably discern some patterns. Some communities, for example, have a strong recreational focus; people gather in large and small groups to play sports, play cards, bowl, and so on. In others there is a large number of volunteers and a large number of volunteer opportunities. In still others, though, people tend to come home from work, lock their doors, and spend time only with themselves and their immediate families.

As you learn about the people in the community, be sure to look for their strengths. In human services, since the focus is often on problems, it is easy to look at people and communities in terms of what they do not have and what they do not offer. And that is an important part of the picture. Most communities also have strengths, however, and those strengths are often found in the skills, talents, and values of the people who live there.

Community Symbols

Communities, just like organizations, can be looked at as cultures. One of the assets of any culture is its symbols and traditions. Earlier in the chapter we mentioned landmarks. Some community landmarks have symbolic importance beyond what is immediately obvious. A church, for example, can be a center of social life in a community. As such, it is a symbol of unity in the community. Communities may have special celebrations, like town fairs, carnivals, dances, or parades. They may also have their own traditions around holidays.

Political Considerations

The issue here is who has power and how decisions are made, and there are both formal and informal aspects to these dynamics. Find out the status and strength of the major political parties in the community. In some communities, there are parties in addition to the Republicans and Democrats that have strong followings. Another part of the formal structure in a community is its local decision-making structure. There are many of these, ranging from small towns that are governed by small committees and a town meeting to large municipal governance structures.

There is more to the political life of a community than decision making, and many more sources of political power and influence than the formal mechanisms. At issue is the control of vital resources, such as money, goods and services, and information. In some cases the institutions that control money, energy, natural resources, and information are located outside the community and operate without any input from or regard for community interests. And don't forget the informal means of access; there are both legal and illegal avenues to loans and credit, for example, and to goods and services.

Sometimes power is conferred formally, by election or appointment, or by occupying a position of authority in a major institution. Other sources of power in the community are less obvious. People in communities derive power and influence from their families, for example. In some communities there are families who have lived there and been powerful for generations. Other people derive power from their connections, their personalities, or both.

Table 9.1 shows all the dimensions of a community that we have discussed. It can serve as a summary as well as a guide for your investigation.

HOW DO I FIND ALL THIS OUT?

First of all, you probably are not going to find out *all* this information about a community, depending on the length of your internship and the importance you, your instructor, and your supervisor assign to these issues. However, everything you learn about a community and about how to do this sort of research will be valuable, and there are several places to look for information. Some you can find out by yourself, by visiting local or college libraries. Most libraries have reference librarians who can help you, provided you can be clear about what you are looking for. Homan (1999a) suggests several resources that you might find in a library, including:

1. Census data
2. City directories
3. Newspaper files
4. Local magazines
5. State yearbooks
6. Town reports
7. Political directories

You may also be able to find a lot of this information on the Internet.

Another source of information is needs assessments of the community, which are frequently undertaken by charitable organizations such as the United Way or by local hospitals, colleges, or universities (Homan, 1999a; Kiser, 2000).

For other sources of information, you will have to go out into the community itself. Local businesses, or large businesses that operate in your community, are a good resource for information. So are chambers of commerce, local human service agencies,

TABLE 9.1
Dimensions of a Community

Basic Characteristics	Structural Issues	Human Resources	Cultural Issues	Political Considerations
Geographic	*Basic Needs*	*People*	*Symbols*	*Formal Power*
Size and boundaries	Food	Sense of safety	Landmarks	Major political parties
Landmarks	Shelter	Sense of belonging	Gathering places	Other parties
General appearance	Clothing	Sources of pride		Governance structures
	Medical care	Core values	*Rituals*	
		Skills	Fairs	*Formal Control of*
People	*Employment*	Talents	Celebrations	Money
Number	Opportunities	Other strengths	Parades	Energy
Density	Training needs			Natural resources
Stability	Location			Goods and services
Income	Unemployment %			Information
	Education			*Informal Sources of*
	Schools			Power and Influence
	Adult education and training			
	Communication			
	Media			
	Informal communication			
	Transportation			
	Air and Water			
	Waste and Drainage			

and other groups that either service or advocate for your community (Homan, 1999a). And if you can find someone at your agency—or anywhere else—that can connect you with some of the informal networks we have discussed, you will learn things about your community that you cannot learn anywhere else.

Be prepared, though, to find out that no one in your agency has these connections. Homan (in press) points out that too often agencies sit *in* a neighborhood, but not *with* the neighborhood, and conduct their business in isolation. In that case, Homan suggests that, as you have time and energy, you begin with clients, merchants, faith community staff—and ask them who else you should be talking with in order to get a better feel for the community.

SUMMARY

We hope that we have aroused your interest and curiosity about the community or communities that form the context for your work. Even if we have not, we hope you will go out and look for some of the information discussed here, and see if you can connect it to the lives and struggles of your clients.

This chapter ends Section Two of the book. We have encouraged you to get oriented, to explore the basic dynamics of your clients, colleagues, agency, and community. In so doing, you have addressed a number of the concerns and issues of the Anticipation stage. With effort, and a little luck, you will feel more settled in your placement and more committed to the work. In Sections Three and Four we consider the challenges that await you as the remaining stages of your internship unfold.

For Further Reflection

FOR PERSONAL REFLECTION
Answer questions 1 and 2, and think about how much time and energy you have for question 3.

1. Think about the clients you work with or the agency that you work for. What sorts of communities have an impact? Include communities of place as well as communities of interest, age, or culture.

2. Which of these communities do you think is most important for you to learn about, and why?

3. Conduct a community assessment. First, plan with your instructor the scope of the assessment and the amount of time you have to devote to it. Next, choose from the aspects of communities listed in this chapter a set that fits the scope of your assessment. Keep track of where you find information, not just what you find.

SPRINGBOARD FOR DISCUSSION
If your classmates are all working in the same community, or if several of you are, you may want to consider taking a quick tour of the community by car or public

transportation.[1] Plan ahead for what you will be observing, and divide up responsibilities. One person might keep track of the number and nature of small businesses, another of churches, another of bars, and so on.

For Further Exploration

Homan, M. S. (1999a). *Promoting community change: Making it happen in the real world* (2nd ed.). Belmont, CA: Wadsworth.

Clear, cogent, and thorough introduction to principles of community development.

Homan, M. S. (1999b). *Rules of the game: Lessons from the field of community change.* Belmont, CA: Wadsworth.

Packed with practical suggestions and accumulated wisdom.

Kretzmann, J. P., & McKnight, J. L. (1997). *Building communities from the inside out* (2nd ed.). Evanston, IL: ACTA Publications.

Good resource for adopting as assets-based approach to communities.

References

Bolman, L. G., & Deal, T. E. (1991). *Reframing organizations: Artistry, choice and leadership.* San Francisco: Jossey-Bass.

Homan, M. S. (1999a). *Promoting community change: Making it happen in the real world* (2nd ed.). Belmont, CA: Wadsworth.

Homan, M. S. (1999b). *Rules of the game: Lessons from the field of community change.* Belmont, CA: Wadsworth.

Homan, M. S. (in press). *Promoting community change: Making it happen in the real world* (3rd ed.). Belmont, CA: Wadsworth.

Horejsi, C. R., & Garthwait, C. L. (2002). *The social work practicum: A guide and workbook for students* (2nd ed.). Needham Heights, MA: Allyn & Bacon.

Kiser, P. M. (2000). *Getting the most from your human service internship: Learning from experience.* Belmont, CA: Wadsworth.

Kretzmann, J. P., & McKnight, J. L. (1997). *Building communities from the inside out* (2nd ed.). Evanston, IL: ACTA Publications.

Rogers, G., Collins, D., Barlow, C. A., & Grinnell, R. M. (2000). *Guide to the social work practicum.* Itasca, IL: F. E. Peacock.

---

[1] Thanks to Mark Homan suggesting this exercise.

FACING NEW FRONTIERS

In Section Two, we helped you prepare for, examine, and work through many different issues that arise in the beginning stage of an internship. Once you are finished with those issues—or at least once they are not foremost in your mind—you can turn your attention to other things. This section of the book is designed to help you keep your progress moving and your internship alive and vital.

In Chapter 10, we identify common traps that can await you at this point in your internship and encourage you to take a careful and thoughtful inventory of the progress you have made and the issues that remain for you to resolve. We also explore a common, although not universal, phenomenon: the feeling that your internship is falling apart; we refer to this phenomenon as the Disillusionment stage. In Chapter 11, we present you with a model for thinking about and dealing with the obstacles and difficulties that you have identified. This section of the text is less theoretical and considerably more applied than the previous one. Our primary objective is to get you to examine your own experience. We believe you already have the theoretical tools with which to conduct such an examination.

Although you may find the reading in this section a bit easier, the issues you will deal with are every bit as difficult as those faced in the Anticipation stage. However, if you have worked through the Anticipation stage, then you know the sense of satisfaction and empowerment that comes with meeting issues head on. In Section Two, you achieved an informed, realistic commitment to your work. In Section Three, you can achieve a clearer vision of that work and an increased sense of confidence that, through reflection, effort, and reaching out for help and support, you can handle whatever comes.

CHAPTER 10

Taking Stock and Facing Reality: The Disillusionment Stage

I knew I would learn an incredible amount because of the actual hands on experience that you cannot get from a book. But I was surprised at the impact the internship had on me and to learn how much I needed to work on.
STUDENT REFLECTION

You have dealt with the issues and concerns that accompany the first weeks of the internship and have moved successfully through the Anticipation stage. Now what? Do you relax and coast through to the end? Well, you might be able to do that (although it's not likely), but you will miss many learning opportunities if you do. Do you wait, apprehensively, for the problems to come? After all, you read about the stages of an internship and know that the Disillusionment stage is the next stage for many. In our experience, almost all interns do experience some problems, but how many, how severe, and when they will occur are difficult to predict. Dealing with them is part, but not all, of the challenges of the next stage of your internship.

One focus of this book is to help you become a proactive learner, shaping your own learning experience as much as possible. So, what we want to talk about now is how you can keep yourself growing and moving forward. There are three major components to this effort: (a) taking inventory of where you are now and where you want to be, (b) identifying any problem areas that need attention, and (c) developing and implementing plans to take action in both of these areas. These steps of assessment, problem identification, and problem solving involve skills that are essential to the success of your internship. You will continue to need and use them throughout your placement. Each of them, though, can involve and evoke powerful feelings.

In this chapter, we begin with some thoughts about growing and learning. We then present some ways for you to take stock of your progress so far. We pose some general questions about your internship and ask you to think about issues and challenges in the major arenas of your internship: the work, the people, the site, and yourself. It is at this time in the internship when many people enter the Disillusionment stage. We help you make sense of that experience and think about ways to move through the challenges.

THINKING ABOUT GROWTH

One of my anticipations was "growth." This I can feel every day. I can feel myself growing as a person as well as a professional.
   STUDENT REFLECTION

The Human Side of the Internship

Why did you decide to do an internship? For many interns, the answer is the same as the reason they are going into their careers: to help people. As time goes on, though, many interns become absorbed in the assessment techniques and intervention skills they are learning. Other internships are focused less directly on people in need, and in those cases, the overall goal may have been skill and career development right from the start. It may seem to you, then, that the overall goal of your internship is to inquire into the nature of phenomena—be they people, systems, or projects—by studying them firsthand in their natural environments and providing necessary services to them; for example, you might be providing services to elders in the community or to adolescents in crises with their families, their schools, or the law. And of course, you want to do the best possible job you can in providing such services.

However, we remind you that your internship is fundamentally a *human* experience (Georges & Jones, 1980). When the focus of internship is on collecting data, analyzing information, providing services, or even acquiring skills, rather than on *living the experience* of the internship, the effects of the human side of the experience on you and your work are easily ignored or may only be mentioned in passing, as if they were of little importance to the real work of the internship. Ironically, it is precisely this human element that makes the field experience one of mutual learning and vulnerability for you and for the others who are part of your experience in the field (Georges & Jones, 1980; Wagner, 1981).

So, how do you make sure you are still attending to the human side of the internship? First of all, you need to remain open to and accepting of your feelings, even those messy ones that you wish you didn't have or think you're not supposed to have. The more you are open to your own emotional experiences, the more you will be open to the emotional experiences of others. Your feelings are also valuable cues to learning opportunities. Second, you need to remain open to the relationships you have in your internship and the joys and challenges they bring. Finally, you need to remain open to individuals. Remaining open means listening, reflecting, paying attention, and resisting the temptation to put people—including yourself—in categories and boxes that limit your experience of them.

Experiencing Change

> *The setting may change but I continue to be who I am for the most part while I can sense some change within myself. Others are who they are. The field experience is rich because what may seem familiar experiences are reviewed with a different perspective. . . . They are getting a twist and that is good.*

> *Conventional wisdom says work smarter, not harder. Word of the day— resiliency. . . . Feel the emotion, bounce and continue. Change is happening.*

STUDENT REFLECTIONS

Moving on and changing are generally accepted as integral parts of the internship process. You expect them, and so do your site supervisor and your instructor, so much so that if they do not occur it can be cause for concern. *Moving on* suggests taking on additional responsibilities that further test and develop your skills and competencies. You expect your workload to increase and your performance to improve. Essentially, you seek out greater challenges and strive to meet them, often with considerable satisfaction.

Changing, on the other hand, suggests that an internship affects you in noticeable and pervasive ways. During the first seminar class, just after field placement begins, we often ask our interns this question: What would you say if I told you that the person you are today, sitting here in class, is probably not the person you will be when you complete your internship? Our interns tend to look at us very strangely and quietly when we pose this prediction to them. How about you? What is your reaction to such a prediction? Would your reaction be different if the question had been asked at the very beginning of your placement?

Maybe the question strikes you as a little dramatic, and perhaps it is, but change involves something more substantial, more far-reaching, and more challenging than just taking on new responsibilities or improving your skills. It involves looking at yourself, your clients, the helping process, the placement site, other organizations, and even society in new ways, and this is not solely an intellectual process.

Change of this sort can be enormously exciting, and it can also be frightening. There are natural tensions involved in such change, and you are bound to feel them. According to Jon Wagner (1981), you deal with two sets of perspectives during your internship: the perspectives you bring to the field, and the ones you are developing as a result of your work. On the one hand, you are drawn to the reality of the work, its vitality, its unpredictability, and its dynamic yet problematic character; on the other hand, you seek the comfort of "home," where you know the situations and the problems and where you are known to others. Your challenge is to bring these two sets of perspectives together in such a way that learning goes on. If you keep both your feet safely planted at home—concentrating only on how you currently think and what you already know— you underextend yourself in the experience and risk not growing; however, if you bury yourself in unfamiliar soil, you overextend yourself and risk sinking in the experience.

Resolving these differences requires that you embrace both perspectives at the same time, "straddling them" in Wagner's (1981) words, so as to create a perspective or vision that incorporates the past as well as the emerging ways of looking at things. Sounds great, right? It can be, but William Perry (1970) reminds us of two opposing human urges: the urge to progress toward maturity and the countering urge to conserve. These competing emotions accompany the two perspectives just discussed and can interfere with their integration. Similarly, Robert Kegan (1982) believes that all change involves letting go of old familiar ways, and that can be frightening. So, expect some excitement as well as some trepidation, and remember that both these emotions can help you grow.

Given these challenges and emotions, how can you keep yourself moving forward? For one thing, you can remember the experiential learning cycle described in Chapters 1 and 3. Mobilize your own resources and those around you, so you can seek and have new experiences. Remember, though, that action and experience are only part of the picture. The need to *do* is very high for most interns, as you have probably discovered, and the needs of placement sites can also be high due to the sheer volume of work and reduced staff size. However, the activities of processing and reflecting are equally critical to your professional development and learning. They need to be structured parts of your internship experience in activities such as journal time and discussion groups with peers, as well as designated personal time for thought and reflection. When the balance between action and reflection is compromised, so is the overall quality of your field experience. How are you doing so far in your internship at achieving this balance?

You also need a balance of challenge and support from yourself and those around you. You need support for who you are now and the way you think and feel, but you also need to be challenged to move forward and take risks (Kegan, 1982). You need support as well as challenge as you experience the emotional dimension of the internship, so you can use that dimension to help you learn and grow.

TAKING STOCK OF YOUR PROGRESS

With thoughts about learning and growth in mind, we turn to the process of assessing your progress. This process involves considering some general questions as well as some specific issues. One theme you will notice throughout this part of the chapter is the reaction you have when something you have read or heard about appears in front of you, or within you, in real human terms.

Keeping the Contract Alive

Remember your learning contract? When was the last time you looked at it or thought about it? In Chapter 5, we described the learning contract as a living, changing document, but we also know how easy it is to set it aside and forget it once it is handed in. The time has come to take stock of this document. Have discrepancies developed between what you planned and what you are doing? Have some of the goals been reached and now seem uninteresting? Are there activities you know you will not get to do? Have there been changes in agency personnel, needs, or priorities? If so, perhaps an activity that once seemed perfectly possible must now be foregone or drastically modified. Have new and interesting opportunities presented themselves? If so, they may need to be incorporated into the contract. If your learning contract is to serve as a valuable guide and point of mutual clarity for you, your site supervisor, and your instructor, you all need to work to keep it from becoming obsolete.

Expanding Your Knowledge

Some of your goals were knowledge goals, and you want to be sure you are challenging yourself, and being challenged, to acquire more concepts and think at new levels. The exact nature of these knowledge challenges is, of course, an individual matter, but there are two general issues to think about.

First, the internship is a tremendous opportunity to integrate theory and practice. You probably have studied lots of theories in your classes, and even given some thought to their application. However, as you leave the role of observer and become a more active participant, your theoretical knowledge changes as the result of trying to apply it (Benner, 1984; Garvey & Vorsteg, 1992). This process can be troubling; you may find that your theories don't seem to work. Could all of your professors and textbooks have been wrong? Not at all! The integration of theory and practice is a normal challenge in the helping professions, because not all theories work equally well in all situations. Furthermore, it may take you a while to recognize your theories in action because they look different when acted out by real human beings with real problems. Ask yourself, How has your experience so far expanded and challenged your theoretical base? If your old frameworks are not working, perhaps you need to request some readings to help you find more suitable ones.

Second, you need to develop some general principles to guide you in your work. These principles are derived from your experiences in class, in the workplace, and in other field and life experiences (Benner, 1984). Often, interns can tell us about what they did in a certain situation and why, but they cannot answer general questions such as: What are you thinking about as you approach a youth in crisis? or, How do you know when to press for a response and when to let a client be silent? You might be able to quote textbook answers, but do not feel a *personal* connection to those answers because they are not coming from your experiences. As you process your experiences, try to formulate some general ideas about what works for you, what works in an agency like yours, and what your strengths and weaknesses are in a given situation.

Expanding Your Skills

When we discussed learning contracts, we mentioned that there are three levels of learning. You are moving into the second phase of learning now, where the focus shifts from learning about the work to learning how to do the work. When you wrote your contract, you probably wrote about skills in fairly general terms. As you begin to acquire new skills, you can think about skill development in more specific terms and specify smaller steps for yourself.

Dreyfus and Dreyfus (1980) found that students go through five levels of skills acquisition: novice, advanced beginner, competent, proficient, and expert. In addition, they found that these levels of proficiency reflect changes in three aspects of skills performance: (a) a shift from reliance on abstract principles to past experiences, (b) a change in perceptions from seeing a series of parts to seeing things as wholes with relevant parts, and (c) movement from being a detached observer to being an involved learner.

The Dreyfus model also has been used to study how health-care professionals acquire and develop the skills to carry out their work effectively (Benner, 1984). Benner's findings are particularly useful when thinking about this new phase of your internship. As a novice, you had little understanding of the situations in which the work must be carried out. The basic skills that you brought to the field were generic and in need of *contextual meaning*, that is, knowledge of how to use your skills effectively in a particular situation. For example, knowing the names of the resources in a community and being familiar with the indicators of potential suicide do not necessarily mean that you know how to interpret a client's behavior and prioritize your responses to deal with a life-threatening crisis.

By the end of your internship, you might be expected to develop the skills to respond effectively to just such a crisis. Certainly, it would have been helpful to take a course to prepare you for understanding the contextual meaning of your work, but that may have been neither practical nor available. At this point, you may be part way down the road to contextual meaning.

CONSIDERING THE ISSUES

Just as there were issues that most interns face in the Anticipation stage of the internship, there are new issues that emerge as that stage begins to fade. Some interns report that they encountered all of the issues, a few report encountering none, and most fall somewhere in between. The issues arise in the broad areas of the work, the people, and the site.

Issues with the Work

Work issues are those that center on the demands being placed on you, the responsibilities you are undertaking, and the opportunities you are being given. There are two issues for you to consider. First, there is the issue of the challenge inherent in the work you are given. Are you feeling overchallenged? Underchallenged? If so, you want to think about the volume, the depth, and the scope of what you are asked to do. Too much responsibility can be overwhelming and leave you feeling ill equipped to do the work; too few demands result in being underchallenged, which can be humiliating, frustrating, and disappointing (Burnham, 1981; Royse, Dhooper, & Rompf, 2003).

Second, you may be doing tasks that are not in your contract. Sometimes exciting opportunities arise. The agency receives a grant, your supervisor learns of a new support group that is starting, or you fill in for someone in an emergency and discover you like what you are doing. There is nothing wrong with these changes as long as you keep your goals in mind. If those goals are changing as well, then a shift in activities makes sense. Do not be lured by the novelty of an activity or responsibility, though, that does not really fit in with where you want to go and what you most need to learn. Sometimes you get these different responsibilities because the agency needs help and you are there. Of course you will want to help, and you will be nervous about saying "No" to your supervisor. But are these responsibilities a good match with what you most need or want to learn? If so, they need to be incorporated into the learning contract. If not, they need to be discussed in supervision.

Balancing your needs as a learner with the agency's needs for an intern can be a challenge. The balancing act becomes even more complicated when you are employed at your field site or are offered employment while interning. Potential problems can be managed if you clarify your needs in both roles by discussing them with your supervisors. It is wise to keep supervision separate for these two roles. This usually means two different supervisors, one for the internship and one for employment.

Issues with the People

For most interns, the internship unfolds in the context of a web of relationships. At this stage, acceptance concerns have usually been addressed. The relationships are continuing to develop, though, and new issues will undoubtedly present themselves. There are issues with clients, with supervisors, with coworkers, and with peers.

THE CLIENTS

If you are engaged in direct work with clients, a great deal of your psychological, intellectual, and emotional energy will go into this aspect of your internship. As you move from an intellectual way of understanding clients to emotional involvement with them, you are bound to have reactions to them and their situations. The challenge here is to understand your reactions and learn from them rather than letting them interfere with your work and your general well-being.

Reacting to Life Situations As you get to know your clients, learning about their individual and collective life situations can be emotionally demanding. For example, if you have already had your first encounter with issues of abuse, disease, or unnatural death, you already know that the work is never easy with clients who are so socially vulnerable (Brill & Levine, 2001; Wilson, 1981). Learning to understand their viewpoints can be challenging and even threatening to your perceptions. So, too, is dealing with and understanding their behaviors. Destructive, irrational, or violent behavior directed at peers, family, or loved ones can be difficult to hear about or see. If the behavior is directed at you, as it might well be if you are interning in the criminal justice system or in a psychiatric setting, the behavior can be most unsettling. (See Chapter 6 for discussion of personal safety issues.)

As you come to know your clients and their situations better, the issue of empathy can surface again. In Chapter 6, we discussed the problem that some interns have finding common ground and establishing empathy with clients. If you are successful, though, you face a new set of problems. Understanding others' viewpoints about and experiences in the world challenges your own, and such challenges can be threatening at times. For example, one intern from a middle-class background was working with juvenile offenders. She could not understand how anyone could commit crimes and think the behavior was right, although she understood that people make mistakes or act impulsively. As she listened to clients, though, and came to understand how hopeless many of them felt and how angry they were at a system that never seemed to benefit them, she began to understand their contempt for the legal system and the reasons for their own code of ethics. She was quite rattled to discover that something she thought she knew for sure was not as certain anymore.

Your emotional reactions to this work can become problematic when they interfere with your learning or with your ability to carry out your field responsibilities. If you find that you cannot get the clients out of your mind and cannot concentrate on other things, a problem exists. "I'm overwhelmed by the sad stories of their lives; I can't get them out of my head. This is too much for me." We typically hear comments like this from students who work with violence and other forms of extreme emotional content. These interns tend to develop needs different from those of their peers early in their field experiences. For example, they tend to need more airtime in seminar class. They also tend to be more cynical in their perspectives and preoccupied with loss, pain, and violence (King & Uzan, 1990). We work with these students between seminar classes as needed and help them to understand how the work is affecting them.

Reacting to Client Progress   If you have developed a reasonably comfortable relationship with your clients, you may be wondering how and when you can move beyond what many interns call "chitchat" into some real work. This concern is not limited to those who work one-to-one with clients. Interns at sites such as shelters, soup kitchens, or drop-in centers often report that clients are now accepting of them. They have stopped giving them funny looks or ignoring them and will sit and talk with them. However, as one intern put it, "When I try to talk about the jam he got into the night before, a difficulty at home, or his drinking problem, he changes the subject. If I keep it up, he shuts down, or just gets up and leaves."

As you spend more time with and around the clients, you may find them engaging in all sorts of behaviors that seem unproductive or even counterproductive. For example, they may refuse to talk; they may talk on and on about their problems but show no interest in solving them; they may agree to try new behavior and then not follow through. Upon examination, there are many ways to understand such actions. Initially, however, interns who are not making the progress they think they should with clients are often frustrated and concerned.

Reacting to Boundary Issues   Another area of reaction has to do with setting and maintaining boundaries. Appropriate boundaries are very important in a helping relationship, although clients may not always appreciate them. When you set boundaries, you are making your clients aware of the limits of your relationship and what is and is not permissible in that relationship. Clear boundaries will help your client understand your role and what to expect of you. They also let the client know where in your relationship there is flexibility. Ultimately, they help the client feel safe.

Clients often test boundaries, and that can be hard on the intern. Clients may ask you for personal information that seems both irrelevant and inappropriate. They may ask whether they can call you at home or ask you to stay with them beyond your shift or the time limit of their appointment. They may invite you to dinner or to visit their home or to attend a family wedding (perhaps even their own!). You can probably think of many more examples and may even have experienced some in your placement. One intern reported, "I realized this week that the staff and I are not at this facility to be friends with the clients. . . . Our main goal is to assist them in their recovery. . . . We must draw the line at a professional level. . . . We must be careful not to get too close. . . . (I'm) aware of (the) trap before I get sucked into it. I imagine that once a staff member gets sucked in, it is difficult to get out." If the words of this intern are familiar to you, then you need to bring the matter to supervision immediately. The trap is very difficult to extract yourself from. Even seasoned professionals find this trap to be seductive at times.

Some boundaries are very clear and unquestionable, such as sexual relationships with clients. If a client is attracted to you, even if the attraction is mutual, there must be no sexual relationship. This is a clear boundary. Other boundaries are more flexible and situationally determined. For example, the issue of socializing with clients can be a complex one, especially if your services are delivered in the community (as opposed to in an agency) or your clients are from a culture that expects some social contact with helpers. Agencies differ in their norms and policies. By now you should be aware of the ones at your placement, and if you are not, it is time to ask about them.

Reacting to Clients   You may be surprised to discover that there are clients and behaviors you just do not like. You may even feel frustrated and refer to these clients as "difficult." The term *difficult client* implies that the difficulty rests solely with the client. While it is true that certain behaviors, behavior patterns, and issues would be troubling to anyone working with the clients, the depth and breadth of the emotional reactions to the client vary greatly from person to person. The question is not which clients are difficult, but which clients *you* find to be difficult. Two studies conducted in the 1980s found similar results in surveying helpers about their perceptions of stressful client behaviors. Suicidal statements, depression, anger/aggression/hostility, apathy or lack of motivation, and the client's premature termination were most stressful to helping professionals (Deutsch, 1984; Faber, 1983).

Reacting to Diversity   Perhaps you are working with clients whose race, ethnicity, age, social class, lifestyle, stage of identity development, or any other aspect of their cultural identity is new to you and very different from your own. You know how you *want* to feel about your differences, but it is important to be open, at least with yourself, about how you *actually* feel. And what you probably did not expect to deal with are your reactions as you learn about people who are quite different from what you expected. For example, there are the sex-abusing clergy, the substance-abusing physicians, and embezzling attorneys. We have met them in our work, and you will meet them, too.

THE SITE SUPERVISOR

You may find yourself feeling troubled about issues with your supervisor, whether it's your impressions of the supervisor, the supervisor's style of supervision, or aspects of the supervisory process. Issues with supervisors can threaten the foundations of trust and acceptance you are working to develop and in turn leave you feeling vulnerable in dealing with them. If you are facing these challenges, the following paragraphs may prove helpful as you work your way through them.

Reacting to the Supervisor   Some interns initially are so impressed with their supervisors that they idolize them. If you find yourself feeling this way, remember that the bubble eventually bursts. The supervisor makes a mistake with a client, snaps at the intern, misses an appointment, or exhibits one or more of any number of foibles that all of us can fall prey to at one time or another. And, when that happens, there is a real letdown. For example, you are apt to become confused by what seems to be callous and cold behavior on your supervisor's part. However, if you do not know that workers who have to deal with heinous acts of violence as part of their workday use outward appearances of insensitivity and callousness as ways of coping, then you might become angry and disappointed with your supervisor.

Reacting to Supervision Style   By now, you know a good deal about your supervisor's style and your own supervision needs. You have a sense of what is suited to your needs and what is not. And, you probably are aware of how your own needs have changed since the internship began. All these factors can affect the supervisory relationship for better or worse. This is a good time to think about how your emerging needs match up with different styles of supervision. You may want to review the discussion of supervisory style in Chapter 7 and what you need to do to get your needs met. One intern decided to handle style issues in this way: "I have learned a lot about her style. I have had to change my style in some ways to meet hers. I know that she runs things in a certain way, and I have finally learned to make my needs match her style. I make sure all of my needs are met, but I make sure that they are on her grounds, so to speak."

Even if you and your supervisor are well matched, you may have had differences of opinion about clients, strategies, projects, or other issues. Acknowledging and discussing differences are considered good practices in the helping professions. However, you may feel anxiety in doing so with your supervisor, especially at this stage of your internship (Wilson, 1981). In most cases, open discussion strengthens the supervisory relationship as long as your supervisor responds empathically and educationally to your concerns. If you are not ready for an open discussion, you may find this intern's comments helpful: "I wouldn't say that we have disagreed, but I would say that I have disagreed with her and not shown it. I dealt with it in my own way by means of reflection and discussion with peers. Then I realized how to fix it without stepping on her toes."

Reacting to the Supervisory Process   There is the potential for issues to arise in the supervisory process as well, especially in the areas of feedback and evaluation. Some people accept both positive and negative feedback gracefully. Others are uncomfortable with it; they become defensive about negative feedback or give someone else the credit for their work when they are complimented.

By this time you have probably had an evaluation. This can be a time when you receive considerable affirmation about your performance as well as your potential as a professional. However, for many interns, this is also a period of introspection, self-doubt, and self-confrontation, which can raise painful issues. For example, your limitations and gaps in learning, skills, and style are identified openly during evaluations, and you may become preoccupied with your shortcomings, especially if you receive more negative feedback about your personal style than about your professional skills. Evaluations may cause you to scrutinize your skills, abilities, commitment to your field of study, and even to question the value of the work itself and your choice of profession (Lamb, Baker, Jennings, & Yarris, 1982).

Reading your evaluation is a little like getting back a paper. Some students read all the comments because they want to learn and grow regardless of their grade. Others look to see whether they got the grade they wanted. If they did, sometimes they don't even read the comments; if they did not, they immediately become defensive and upset.

We hope you can take the former attitude. Even so, you may have a variety of reactions, whether positive or negative, including speechlessness, disbelief (some students cannot believe how good their evaluations are), anger, and tears. Some students who have gotten As in their classes may be shocked at being critically evaluated for the first time. It may be that they did very well, but the supervisor also noted areas for improvement. It may also be that they are better in the classroom than in the field, at least at this point. Some interns who are doing a second internship and received high ratings previously are surprised if their current ratings are not so high. Perhaps the previous supervisor was unwilling to be critical, for any number of reasons (Wilson, 1981). If there are other interns at your placement, especially if they are from your campus, there may be some subtle, or not so subtle, competition concerning evaluations. If you think your supervisor was harder on you than on a peer, or if your peer had another supervisor whom you think is an "easy grader," you may have trouble staying focused on learning from your own evaluation.

Many interns are reluctant to question an evaluation, but if you don't understand why you received a rating or comment, it will be very difficult to learn from it. Keep in mind that you have a right to a high-quality evaluation, and if you do not receive one, you should consult with your instructor. *High quality* does not necessarily mean favorable; we have both seen very positive evaluations that we thought were poorly done. Suanna Wilson (1981) suggests that a high-quality evaluation is concise, specific, describes behavior (as opposed to using labels), and contains both positive comments and suggestions for improvement.

THE FIELD INSTRUCTORS, COWORKERS, AND PEERS

In many ways, the same issues that arise with site supervisors arise with the field instructor who is overseeing your placement and/or conducting the seminar class. You do not spend anywhere near as much time with this person as you do with your site supervisor, but it is an important relationship nonetheless. This relationship can also have the added dimension of grading; in most programs, the instructor is partially or wholly responsible for issuing grades. Now is a good time to reflect on how that relationship is going.

There are four sets of major issues that arise with coworkers. The first set of issues has to do with the tendency of other staff to give work to you or otherwise exert influence over you, usually without your supervisor's awareness. The second set of issues concerns coworkers who seek your friendship and support. Coworkers may have hidden motives for befriending you, and it may take you some time to figure out what is going on. Often, these members of the staff are at most marginal and most likely to reach out for alliances. In addition, if there are factions at the site, people may become friendly in order to recruit you to their side or clique. The third set of issues has to do with becoming part of the staff team. Interpersonal conflict with one or more coworkers or the general feeling that you are not a team member can be major source of stress for interns (Yuen, 1990).

The fourth set of issues centers on the kind of example that coworkers set for you. Some of what you have seen from your coworkers probably warms your heart and gives hope for your future work. However, perhaps not all of what you have seen felt right or reassuring. For example, perhaps you overheard some workers coolly discussing the murder of a child, which you recall from the morning paper. As the conversation went on, you realized that the child was a former client of the agency. Other than this conversation, it was business as usual, and few outward emotions were evident. Again, if you were not aware that outward appearances of insensitivity and callousness are ways of coping with the violence that is part of their workday, then you might have become very angry with the situation and very disillusioned about the people who work in the field.

> *Before beginning the internship, I had these ideas about what the people would be like, what the procedures would be like. Some of those expectations were met and some blew my mind. It was the reality of fair-does-not-equal-justice, and the judge-does-not-equal-unprejudiced....*
> STUDENT REFLECTION

It may be that you have seen some behavior for which there is no good explanation. Just as in any field, there are human service workers who are lazy, jaded, harsh, or unethical. Such circumstances can leave you feeling quite alone in your reactions. In addition, the support you need may not be forthcoming from the staff. All of these issues are common to internships and should be addressed in supervision as early as possible in the placement.

Your peers are your classmates and other students who are interning at your site. These are the people who know best what you are experiencing. There are three issues that interns face with their peers. The first is support. Peers can be an enormous source of support throughout your field experience. On the other hand, they can also let you down by talking too much (or too little), offering unwanted advice, focusing only on their issues, or simply appearing indifferent.

Second, same-site interns face the unique issues of rivalry and competition. The rivalry can emanate from either the staff or the interns and usually results in interns being labeled winners or losers. Competition in and of itself is not unhealthy nor necessarily should be discouraged, as everyone may benefit from it. However, when competition becomes the focus of the interns' energy and each intern is watching the other's assignments, level of support, progress, and involvement with staff, difficulties can develop.

A third issue that creates challenges for same-site interns is differences in progress. There are lots of reasons why placements turn out as they do. However, when there are two interns at the same agency and they are having very different experiences, the peer relationship can be affected. No longer are the interns sharing similar experiences, and it is difficult to keep both involved in supporting each other. Such differences in progress need to be addressed as potential issues by you, your supervisor, and your instructor.

Issues with the Site:
The Organization and the System

As with other aspects of your work, seeing the dimensions of organizational life unfold can be quite different from studying about them or discussing them in abstract terms. What an agency values, for example, is evident in how the work gets done. What you see, though, may not always reflect what the agency says it values in its mission statement, and you may find yourself having a value conflict with the agency and the system. This is often the case with the paper-versus-people issue. Because there is never enough time to give clients all that they need and attend to the required documentation of the work, tension develops. "People before paper" may be what you value, and it may be what the organization wants to value, but ideals get put aside because of the paperwork they create. Value conflicts such as this can be disheartening (Sparr, Gordon, Hickham, & Girard, 1988). Once you come to appreciate how these discrepancies develop, you are prepared to be part of the solution.

Second, the agency and system have found ways to organize their resources to get the work done, both formally and informally. The informal organizational structure, or the *structure of influence*, can be a source of issues for an intern because it is this structure that influences what *really* happens in the course of the agency's work (Stanton, 1981). Sometimes these informal networks function quite smoothly and support the overall goals and work of the agency. Other times, they seem to undermine those purposes. Although you have read about such discrepancies, actually seeing them operate and feeling their effects can leave you feeling confused at best and disoriented at worst. Ignoring this reality is sure to frustrate you continually in your work.

Third, you may be starting to feel the effects of the agency's *operational* philosophy—not the one described in the policy manual but the one that has evolved to get the work done. Many interns are impressed with their agency's operational philosophy, though there is also room for philosophical and political debate. On the other hand, you might find yourself frustrated by what you consider to be the coldness and slowness of the system, the reduction of people to inhuman status (consumers, customers), and the realities of dealing with bureaucracies and underfunded and understaffed programs.

Issues with Yourself

We have been encouraging you to take stock of your evolving relationships with your work, the people, and the system. However, we have not paid much attention to another important factor in the relationship equation: you! Finally it is time to take a look at what you are discovering about yourself. Interns consistently tell us that they learned a tremendous amount about themselves during their placements. "It's a power trip in understanding me" as one of our interns recently said.

ASSESSING YOUR MEANING MAKING SYSTEM

Your meaning making system consists in part of your attitudes, values, behaviors, unresolved issues, and psychosocial and cultural identities. Combined, they create the filters or lenses through which you look at your life and make sense of your experiences. Typically, these become more alive as an internship progresses. For example, if experiential learning is new for you, you may have discovered that your learning styles and preferences differ from what you experienced in traditionally taught classes.

It is important that you take the time to develop your reflective skills, for these are skills that will keep you in touch with your meaning making system. Reflection, as you know, is an important part of the experiential learning cycle. It is also an important part of the work you do. Donald Schon (1983) maintains that reflective practitioners reflect at deeper levels of thinking, taking a variety of factors into consideration every time they make a decision. Reflective workers think about what they do before they do it, while they are doing it, and after they have done it. They take time to assess that system by reviewing their decisions, noting what factors went into them and what factors were ignored. You have made enough decisions in your placement that you can begin to form some ideas about your meaning making system. Perhaps you recognize some of these factors as part of yours.

BUMPING INTO YOUR OWN ISSUES

Unfortunately, you cannot predict with accuracy how you are going to think, feel, or react until you get into a given situation. Inevitably, there will be differences between what you thought would happen and what actually happens. For example, it is one thing to think about your values and what it may be like to have them tested or be involved in a value conflict. It is another thing to experience the often unidentifiable feelings that conflicts in values create. This intern's reflections are dealing with just that: "Once I got somewhat settled in . . . I began to realize that it's not all 'up,' not everyone is nice, and working this type of schedule makes me very tired."

Your experience at the internship can also teach you about aspects of yourself that you have missed up to now. You may not know you hold a value strongly, for example, until it is tested in some way. Or you may find yourself reacting strongly to a situation and not knowing why. After thinking about it and discussing it with your supervisor, you discover that the situation tapped an unresolved issue in your life.

Such differences can create just the climate for your own reaction patterns to be become problematic. You will need to label those patterns and responses so that you can make sense of the issues you are experiencing. This student did and came to an important realization: "I would have to say that the biggest and most important thing I have learned in all areas is that I am (only) one person. I am not helping myself or my patients by trying to please everyone. I learned that by spreading myself too thin, I can only hurt the patients and myself, and really not accomplish anything."

FINDING MATCHES AND MISMATCHES

If you have not yet done so, now is a good time to think about what you have learned about yourself and the areas of matches and mismatches in the internship. Perhaps you feel as this intern did: "I am in a constant struggle between my own emotions. When I am given something to do, I feel useful and productive. When I am left alone to find something to do, I feel useless and out of place, and unproductive." This intern's inability to self-direct through the tasks and involve himself in tasks that have meaning is creating an existential problem for him.

In another situation, an intern working at a family planning clinic regularly fielded telephone calls asking for information on abortion. Over time, she became impatient with the phone calls, the clients, and even the agency. She said that many of the questions she answered were "stupid" and wondered why the people were so poorly informed. She chafed at the bureaucratic procedures employed by the agency. Upon further exploration, it appeared that a values issue was at work. The intern was firmly opposed to abortion for herself and had been open about that belief with her instructor and the agency. However, she insisted that she believed with equal fervor that everyone should make their own choices and that she was not opposed to abortion in general. However, as the semester progressed, she grew more and more uncomfortable with the fact that abortions were performed at her agency; it seems that she had much stronger anti-abortion beliefs than she thought. It took her some time to admit this change, even to herself, because she believed she should be tolerant. Once she did so, she was able to direct her anxiety away from clients and the agency and view it as a poor match, given what she had learned about herself. The agency arranged for her to work in their community education division, and the placement was ultimately successful.

There are many other patterns and issues that might cause you some difficulty depending on the circumstances of your placement. Here are some more examples:

- You have dealt with the same issues as your clients, and you are not as "over it" as you thought.
- You have a hard time accepting criticism.
- You say "Yes" when you want to say "No."
- You struggle to establish responsible relationships with supervisors.
- You are extremely upset by confrontation.
- You have a need to smooth over conflict.

We encourage you to identify these and other issues that have arisen in your placement and to share them with your instructor, peers, and supervisor.

CONSIDERING YOUR LIFE CONTEXT

You have been doing the juggling act for a while now, and more balls may have been thrown your way than you can manage. As you are learning, life does get in the way of your internship... illness, a lost job, an accident. You may find your energy level waning, or you may be bending under the strain of multiple demands. You may even feel as this intern did: "I have no time to be myself, or to be by myself, never mind spending time with friends or family." And, spending that time is very important. Friends and family are central to your support system and your sense of well-being. Just as you cannot really know how emotionally demanding the internship will be, you cannot know how well your support system will respond when it comes to internship issues. It is time to take stock of your support network and decide what changes if any are needed.

FACING REALITIES

Sooner or later, you will face some problems with your internship. We are not talking about something minor that is quickly resolved. We are referring to situations that are more substantial, troubling, and that stay with you for a while. We cannot predict how many problems you will have, what they will be, or how you will react to them. Some interns experience only minor upsets; others have a more profound period of upheaval that causes them to question themselves and their placements. Whatever your experience is or will be, we hope this part of the chapter helps you make sense of what is happening and guides you to approach it in the most growthful way possible.

Problems can happen at any time in an internship; perhaps you have already experienced some. If you are several weeks into the internship, it is a likely time for problems to emerge. For some of you, the initial glow of the internship has faded, and some of the aspects that you do not like are becoming clearer. After you have settled in, people often expect more of you, and that can be hard. This intern's stellar performance in the first five weeks of her placement resulted in her being well over her head with the volume of responsibilities she was willing to take on. She needed to put the brakes on and regain control of the internship: "I feel as if they will look down upon me (if I slow down), and I will get a bad grade. Plus, I don't know what my priorities are. . . . It's making me stressed. There are too many things and I have no focus."

One common source of problems is the gap between anticipation and reality. There is almost always a difference between what you anticipated about the internship and what you actually experience in the field. Fortunately, if you acknowledged and dealt with your concerns in the Anticipation stage, you will be less likely to have a wildly different reality from what you expected. However, there will be some discrepancies between your expectations—about the people, the work, and yourself—and the realities of the placement, some of which may be troubling. It may be that the issues and personalities are not what you expected or that you are reacting differently than you thought you would. In addition, issues will arise in placement that you simply never considered or never knew existed because you had no way to anticipate them. These issues derive from that dimension of knowledge about which you don't even know you don't know . . . if you know what we mean! One student observed, "What I don't know I don't know changes all the questions."

Problems can be small or large, but the first one you have usually has an added impact, precisely because it is the first one. You may not have expected to have problems, or at least not the ones you are now having. As we have pointed out many times, experiencing something is different than reading about or anticipating it. Once you have encountered—and resolved—the first problem or two, the experience will be different. Subsequent problems may be harder to resolve, but the shock of the first one or two will be behind you.

The sources of problems are as varied as interns and their placements. Some of the problems that our interns have encountered are listed below, written in the language of their student journals. Perhaps some of them look familiar to you. Perhaps you can add some of your own:

- I really can't accomplish much with these people [clients]—too much damage has been done and I can't perform miracles.
- I've never experienced what they have—they are writing me off. I can't work with them and they know it.
- A client lied to me. Or manipulated me. I thought they trusted me.
- A client had a relapse. Great. Now what am I supposed to do?
- There are some clients I just don't like. I can't find common ground with them.
- They just won't respond to me; not all of them but some of them.
- I have a good surface relationship but I can't move beyond that to really challenge them and explore some issues.
- The pace here is totally insane (or incredibly slow).
- My supervisor is too vague (or awfully blunt).
- My supervisor never seems to have time for me.
- My supervisor is always looking over my shoulder—he doesn't trust me.
- If I don't have any questions, my supervisor ends the supervision hour. That's not right. She's supposed to ask the questions.
- I don't like the way my supervisor treats the clients (or staff).

- The staff are cliquish—I don't belong.
- They dump their busy work on me and send me on errands—now
- I'm a gofer!
- They joke about clients behind their backs.
- They are so cynical. BORING.
- They are trying to get me involved in their problems. I'm just an intern here. I don't need to know what's going on. It's not my problem.
- I'm just an intern. Why does everyone expect so much of me?

Just as there is a range of problems, there is a range of emotional reactions to them. Some interns seem to take the problems in stride; others are really thrown. As you read that last sentence, try not to fall into the trap of saying to yourself, "Well, I am going to be the first type." Perhaps you'll be very accepting and supportive of those who do seem to be thrown, but you won't allow yourself that experience. Remember, allowing yourself to *feel* your problems, as well as to catalog and analyze them, is one way of remaining open to the human experience of the internship. There is no right way to react. Your reactions—and those of your peers—will depend on your emotional styles, the intrapersonal issues touched by a problem or situation, your willingness or unwillingness to be open with yourself and others about your feelings, and, of course, the nature of the problems themselves.

Sometimes students disclose their feelings of disappointment to the faculty, whether by e-mail, conversation in person, or by telephone; other times, they disclose them in the seminar class to their peers. Regardless of how or when the student shares his or her feelings, it is usually difficult for the student because of all the expectations the student had of the field experience and how the student now feels.

In spite of our efforts to reassure them, interns often believe that if they talk about their difficulties or problems, or discuss their concerns or mistakes, then their grade in the field will reflect these shortcomings. It is true that the grade you earn in a field experience is influenced by your strengths as well as by the competencies that need continued development. However, your faculty instructor recognizes that all interns have shortcomings, as do practitioners and faculty. Your willingness to recognize and face the issues helps the faculty instructor gauge your growth in placement. In fact, not doing so may give the faculty reason to question how much you are gaining from your placement (Gordon, McBride, & Hage, 2001).

Most important, whenever you or your peers are discussing feelings of disappointment, which in our experience tends to happen in the seminar class, it is very important that effective communication and feedback skills be used to support the student in need. Knowing when to clarify, paraphrase, and reflect feelings is critical to your and your peers' willingness to come forward again with similar concerns. Failing to use I-statements, being vague and general, and being interpretive when giving feedback is sure to disappoint all who are part of the discussion. This is a good time to re-visit the practices of effective communication (Johnson, 1999) and feedback (Porter, 1982) that we discussed in Chapter 1. You or a peer may soon be feeling disappointment with your field experience, and it is important that you are prepared to support your peers and be supported by them.

Whatever Happened to My Internship?

For some interns, the problems they encounter are troubling aspects of an otherwise positive experience. Other interns reach a point where the whole emotional tone of the internship changes for them, and not for the better. Instead of enthusiasm and nervous excitement, they feel anger, resentment, confusion, frustration, and even panic. You may recall these characteristics of the Disillusionment stage from Chapter 2.

If this description fits you, you are now coping with an enormous amount of the unexpected—not what you planned, but what you are living with during your internship. One of our interns had a long and successful career in a very emotionally demanding field in the helping professions before she returned to campus to complete her degree. She was not prepared for the emotional toll of the internship on her full and demanding life. "I really needed a break. I dropped the ball last week because I was just about cooked to a crisp. This ol' gal was running on fumes."

What you also may find unexpected is the end of the hopeful feelings from earlier days in placement and the onset of more negative feelings. There is the unexpected drop in your enthusiasm and the subsequent drop in your productivity. (Negative feelings do interfere with learning!) Concerns at this time center on many of the same areas as in the Anticipation stage, except that there is a shift in concerns from *What if?* to *What's wrong?*: What's wrong with my internship? What's wrong with my clients? What's wrong with the organization? What's wrong with me?

Early on in your internship, we asked you to think about a metaphor that best described what your internship is like. Now is a good time to take stock of how that metaphor is evolving as you face the realities of the internship. Is the metaphor still working for you? How does your metaphor account for this period of diminished enthusiasm? We have often likened the internship experience to a roller coaster ride, with peaks and dips throughout, perhaps no different from the waves of highs and lows in your daily lives. However, this period is one of distinct character, a dip that leaves students, instructors, and site supervisors alike somewhat perplexed and overwhelmed at times.

The changes brought about by this shift in concerns are often subtle at first, but they persist and begin to become pervasive. You may be complaining to friends or family without even realizing it or hesitating before commenting on your internship to friends or coworkers. You may notice that you are having trouble getting to your placement on time, not looking forward to going, or muttering to yourself under your breath about the situation. Another unexpected consequence of this period is the tendency to direct your feelings at those around you, especially those who are connected to the internship experience—your site supervisor, clients, faculty instructor, coworkers, or even your peers. Sometimes the seminar class itself becomes the focus of your feelings.

The feelings can also be directed inward, and this can be a time when an intern's self-image takes a bit of a beating. For example, when clients do not improve as you expect, refuse to continue services, or react to you very negatively, it can call into question whether or not you were adequately prepared for the work, whether you are cut out to do the work, or whether you should be doing the work at this time in your education (Skovholt & Ronnestad, 1995). In addition, many students entering an internship are used to their friends telling them that they are very easy to talk with and helpful, and the students are often filled with a sincere desire to help. However, there is a difference between wanting to help and actually being helpful and between being an easy person for friends to talk with and being effective with challenging clients. When the inevitable difficulties and criticism come, these students may become resentful, perfunctory in their job performance, or critical of the supervisor (Blake & Peterman, 1985). Perhaps, too, you are not able to work collegially with coworkers or supervisors or within the agency guidelines; perhaps you find little compatibility of values or philosophy with the site.

Earlier in the chapter we talked about interns disclosing feelings of disappointment and frustration with their peers and field instructor, and learning to deal with the feelings and situation through a supportive community in the seminar class. It probably comes as no surprise to you that it is a lot easier to discuss such feelings with peers and faculty than it is with a site supervisor. In fact, it is the site supervisor that interns find most intimidating when it comes to dealing with these feelings in particular. The reasons vary considerably. Think about it for a moment—your site supervisor is responsible for working with you to create and develop your field experience, and right now that experience is very disappointing. Not that you are blaming the supervisor, but it would be very helpful if he or she changed things for you so that you would be more satisfied with your placement. Of course, the site supervisor may not know that you are disillusioned; and, even if the supervisor did know, the changes that are needed are not necessarily something the supervisor can control. Nor is it necessarily the supervisor's responsibility to "fix" your field placement.

The reasons why interns find it so difficult to discuss the Disillusionment stage with the site supervisors vary considerably. Sometimes the issue is one of not wanting to disappoint the site supervisor. Other times it is a fear of the supervisor ending the placement. Chances are the supervisor has dealt with such feelings in his or her own career, but you are not aware of that, so you may tend to think the worst of the situation.

In reality, site supervisors are often a wellspring of resources and support for their students and welcome the opportunity to work with students in resolving such challenges. However, if you are not feeling comfortable in discussing this stage with your site supervisor, and it becomes apparent that you really need to, what do you do? Did you and your peers develop scenarios in seminar class that would be helpful? Has a peer experienced a similar reluctance? If so, how was it resolved? You may want to try what some of our students have done. If the supervisor has been given any information about the developmental stages of an internship, you are well on your way. If such information is lacking, you will need to ask your field instructor to provide you with some. Then, during your supervision time, bring the Disillusionment stage information to your supervisor's attention and mention that you think you are in that stage. Of course, you'll be asked why you think so, so you may want to prepare mentally for that moment, because that tends to be the moment students most dread. Once that moment is behind you, you are ready to confront the issue and resolve the feelings. But that comes later, in the next chapter. For now, it's important to understand the bases for the feelings and your reaction to them.

Earlier in this chapter, we discussed the integration of theory and practice. Garvey and Vorsteg (1992) have suggested that this process can be quite difficult for some interns. It can be a time in placement when interns temporarily reject their previously held beliefs about the theories they work with and experience a crisis in confidence about the worthiness of the work they are doing. It is not unusual for these interns to become disoriented about their learning and appear to go into a funk, experiencing intense frustration, blaming their clients, and questioning their commitment to the work.

Generally, we have observed three potential categories for these feelings of disappointment and frustration. The first is *loss of focus* on the internship. Essentially, you are not doing what you went there to do because the focus of your placement has changed either intentionally or through neglect. Interns who lose the focus of their placements can feel it almost immediately; it is a feeling of being sidetracked or otherwise ignored or discounted. The second is a *loss of accomplishment*. You are not doing what you went there to do, not because the focus changed, but rather because you either are not able to demonstrate the skills or competencies needed to accomplish the internship, or the design of your internship does not allow you to have the experience you wanted. You will probably feel frustrated and angry, feelings that you may direct at yourself or at the personnel or clients at your site. Finally, the third is a *loss of meaning* in the internship. Essentially, you are unhappy with what you are doing in placement. Either the work has no personal meaning for you, or the internship has not been designed to provide you with a meaningful field experience. Regardless, you may find yourself going through the motions of an internship, but without the spirit that comes from being personally invested in the learning and work.

There is no predictable time when these changes in feelings occur; for some interns, they never happen at all. Often, though, these changes are noticed during the phase of learning when the workload increases and there are greater demands for skills. However, the sense of disillusionment can occur earlier in placement, when anticipation concerns are highest, or later in placement, with concerns about competence in the work.

If you are in the middle of this sort of slump, or if you hit one later on, try to remember that it is perfectly normal; in our experience, more interns go through it than do not. Many authors feel that some level of dissonance is necessary for growth. Perry (1970) argues that shifts in ways of thinking about the world are caused by conflict and dissonance. Similarly, Langer (1969) identifies conflict as playing an important role in progress and development and contends that for change to happen, you need to feel that something is wrong. He calls this a state of disequilibrium and says that it is necessary to raise the energy and emotions you need to make changes and put your life back on track. Furthermore, many authors and researchers note that a period of disillusionment is a normal part of any internship (Garvey & Vorsteg, 1992; Lamb, Baker, Jennings, & Yarris, 1982; Suelzle & Borzak, 1981; Sweitzer & King, 1994).

SUMMARY

This time in your internship is full of opportunities. In the early part of this chapter, we discussed the opportunity you have to keep pushing yourself and to keep your internship fresh and challenging. Problems, on the other hand, may seem more like a curse than an opportunity. We certainly feel that way sometimes. However, you will handle your problems better and learn more from them if you take a more constructive view.

Whether you are experiencing isolated problems or a more pervasive disappointment in your internship, facing and resolving the issues confronting you are very important. You might be able to ignore the problems, but chances are that will make things worse. Making sense of the feelings and concerns of this stage is critical to growing through it. And growing through it is critical to a successful internship.

Growth lies in the constructive activity of putting things back in order; resolving the conflict is a prerequisite for progress to the next stage of development and for normal and healthy adjustment to occur. Remember, the new state of equilibrium (feeling that all is well) does not cause progress to occur; rather disequilibrium is the source of growth and development (Langer, 1969).

An interesting and useful perspective on problems and their resolution is offered by Watzlawick, Weaklund, and Fisch (1974). The terms *difficulty* and *problem* are often used interchangeably, and indeed we have done so in this book, but Watzlawick et al. draw a distinction between them. Difficulties, they say, are normal, if unpleasant, events that either have a fairly simple solution or no solution at all. However, in your attempts to solve these difficulties, you may make them worse or bring on additional troubles. It is then, say Watzlawick et al., that you have created a problem. Many of the issues you face as an intern are normal and common. Knowing that may make it a little easier for you, but probably not enough. However, if you ignore or misdirect your energies in trying to address difficulties, they can snowball into full-blown problems.

You may recall that in Chapter 2 we described the Disillusionment stage as a *crisis of growth* and a time of risk and opportunity. If you do not try to face and cope with the problems that arise, you can end up feeling stuck. Those feelings of resignation and resentment will not go away; instead you will suffer bravely (or not so bravely) through them, just as you would a course that turned out to be a disappointment. On the other hand, you have an opportunity to learn more about your work and yourself. More important, you have an opportunity to feel empowered. If you can get through this time, you will feel a new kind of confidence. This is not the false bravado or naiveté that tells you there won't be any more problems, but the confidence that you can resolve them and learn about yourself and others in the process.

Some interns at this point in the placement begin to question their choice of career and their suitability for it. That can be quite unnerving, especially so close to the end of your professional preparation. Questioning your choice of career may make sense, though, and not only because affirmation of the choice is very important. It also makes sense that the choice can withstand the scrutiny of reflection at any time, especially at the beginning of your career, just before you commit yourself to your first position. You may in fact conclude that the specific career you had your eye on is not for you after all; it does happen. Finding that out now, and not in the middle of your first position, is a blessing for all concerned. Furthermore, if you have come this far, you have certainly gained knowledge and skills that are going to be useful to you whatever you do. However, this is no small problem you are facing, and it will need to be dealt with.

The first step in moving through the challenges that face you is to take stock and get a clear inventory and awareness of the obstacles. This intern summarized her style of taking on challenges: "When I am faced with an overwhelming demand, I tend to initially stress myself out with thoughts like, 'How am I gonna do this?' Then I am able to self-talk and make a plan. I typically react by over-reacting and as soon as 'the crisis' is over, I manage and become strong willed or determined." Earlier, we asked you to consider how you were progressing in several areas. Perhaps some issues surfaced then. Perhaps others have crystallized as you read this chapter. The second step, of course, is developing a plan, and that is the subject of the next chapter.

For Further Reflection

Part A: Taking Stock

PERSONAL REFLECTION:
SELECT THOSE QUESTIONS MOST MEANINGFUL TO YOU

1. Reread your answers to the Anxieties and Anticipations exercise at the end of Chapter 5. Have your concerns changed over these first weeks? If so, how?

2. Go back and reread your daily journal entries. What would you say are the major things you have learned so far about your clients, about working with them, about the agency, and about yourself?

3. When it is time to go to your internship, how do you usually feel? Be honest. Has that changed over the course of the semester? If so, what do you make of that change?

4. Take a look at the goals in your learning contract. List them in your journal and indicate whether each has been met, not met, or partially met. In the latter cases, are you disappointed? How might you move beyond these goals to the next logical challenge?

5. What things have happened that you didn't expect? Include positive and negative things and think about the three areas (knowledge, skills, and self) as well as your responsibilities, the client's, and your supervisor's.

6. In general, would you say the tasks you have been given so far have overchallenged you, underchallenged you, or been about right? Explain your answer.

7. Are there clients you are beginning to feel close to? Are there some you have a hard time connecting with? What are your ideas about why?

8. What specific issues do you see for specific clients that go beyond the basic establishment of a relationship with you?

9. What have you learned about your supervisor's style and how good a match it is to your needs?

10. Have you and your supervisor disagreed yet? If so, how did that happen? How did it make you feel? How was it resolved?

11. Have supervisors or others given you verbal feedback on your performance? How did you feel about what they said?

12. How accepted do you feel by your coworkers?

13. In what ways have your coworkers met or not met your expectations?

14. Do you have any concerns about your relationships with your peers—other interns at your site and/or students in your seminar class?

15. How has the agency itself compared to your expectations? Does the pace approximate what you thought it would be? What are the unwritten rules, or values, of the agency? How do you feel about them?

16. Look back at the questions you answered at the end of Chapter 3. Are any of these, or any other intrapersonal issues coming up at the internship? How are they affecting you?

17. How are you doing with managing your other responsibilities? What are the areas of stress or friction?

18. Now that you have been at your placement for a while, does the support system you described in an earlier entry seem adequate? If not, in what ways is it failing you?

SPRINGBOARD FOR DISCUSSION

Take a moment to think about the kind of relationships you have with clients. In what ways have you been challenged by the clients? Have they tested your boundaries? If so, how? What other boundary-related issues have surfaced up to this point? Discuss your answers with classmates and try to help each other clarify the problems and perhaps some solutions.

Part B: Facing Realities

FOR PERSONAL REFLECTION

In light of your accomplishments in the field so far, are there some new goals you want to set for yourself? Be specific.

SPRINGBOARD FOR DISCUSSION

What problems have you become aware of at your internship? Make a list of them and try to put them in order of importance. How does your list compare to those of your classmates in seminar class? What similarities are evident? What differences do you notice?

For Further Exploration

Albert, G. (Ed.). (1994). *Service learning reader: Reflection and perspective on service.* Raleigh, NC: National Society for Experiential Education.

Collection of inspiring and informative readings compiled by the Center for Service Learning at the University of Vermont used as a textbook in field study programs.

Benner, P. (1984). *From novice to expert.* Menlo Park, CA: Addison-Wesley.

Seminal book in clinical practice in nursing looks at five levels of competencies and the demonstration of excellence in actual practice.

Borzak, L. (Ed.). (1981). *Social work field instruction: The undergraduate experience.* Beverly Hills, CA: Sage Publications.

A compilation of articles including students' perceptions of the field experience.

Dreyfus, S. E., & Dreyfus, H. L. (1980). *A five-stage model of the mental activities involved in directed skill acquisition* (Unpublished Report F49620-79-C-0063). Air Force Office of Scientific Research (AFSC), University of California, Berkeley.

Brief, informative paper describes a model of skills acquisition based on a study of chess players and airline pilots.

Garvin, D. S. (1991). Barriers and gateways to learning. In C. R. Christenson, D. A. Garvin, & A. Sweet (Eds.), *Education for judgment* (pp. 3–13). Boston: Harvard Business School Press.

Power of dialogue discussed as part of the process of learning, self-discovery, and self-management.

Ronnestadt, M. H., & Skovholt, T. M. (1991). A model of professional development and stagnation of therapists and counselors. *Journal of the Norwegian Psychological Association, 28,* 555–567.

A model explaining periods of growth and nongrowth in the counselor's professional development.

Rosenthal, R., & Jackson, L. (1968). *Pygmalion in the classroom.* New York: Holt, Rinehart & Winston.

A classic text, illustrating the power of expectations on performance.

References

Benner, P. (1984). *From novice to expert.* Menlo Park, CA: Addison-Wesley.

Blake, B., & Peterman, P. J. (1985). *Social work field instruction: The undergraduate experience.* New York: University Press of America.

Brill, N., & Levine, J. (2001). *Working with people: The helping process* (7th ed.). New York: Longman.

Burnham, C. (1981). Being there: A student perspective on field study. In L. Borzak (Ed.), *Field study: A source book for experiential learning* (pp. 65–72). Beverly Hills, CA: Sage Publications.

Deutsch, C. J. (1984). Self-reported sources of stress among psychotherapists. *Professional Psychology: Research and Practice, 15*(6), 833–845.

Dreyfus, S. E., & Dreyfus, H. L. (1980). *A five-stage model of the mental activities involved in directed skill acquisition* (Unpublished Report F49620-79-C-0063). Air Force Office of Scientific Research (AFSC), University of California, Berkeley.

Faber, B. A. (1983). Psychotherapists' perceptions of stressful patient behavior. *Professional Psychology: Research and Practice, 14*(5), 697–705.

Garvey, D. C., & Vorsteg, A. C. (1992). From theory to practice for college student interns: A stage theory approach. *Journal of Experiential Education, 15*(2), 40–43.

Georges, R. A., & Jones, M. O. (1980). *People studying people.* Berkeley: University of California Press.

Gordon, G. R., McBride, & R. B., Hage, H. H. (2001). *Criminal justice internships: Theory into practice* (4th ed.). Cincinnati, OH: Anderson Publishing.

Johnson, D. W. (1999). *Reaching out: Interpersonal effectiveness and self-actualization* (7th ed.). Boston: Allyn & Bacon.

Kegan, R. (1982). *The evolving self: Problem and process in human development.* Cambridge, MA: Harvard University Press.

King, M. A., & Uzan, S. L. (1990). *The field experience: An integrative model.* Workshop presented at the annual conference of the National Organization for Human Service Education, Boston.

Lamb, D. H., Baker, J. M., Jennings, M. L., & Yarris, E. (1982). Passages of an internship in professional psychology. *Professional Psychology, 13*(5), 661–669.

Langer, J. (1969). Disequilibrium as a source of development. In P. Mussen, J. Langer, & M. Covington (Eds.), *Trends and issues in developmental psychology* (pp. 23–37). New York: Holt, Rinehart & Winston.

Perry, W. G. (1970). *Forms of intellectual and ethical development.* New York: Holt, Rinehart & Winston.

Porter, L. (1982). *Giving and receiving feedback: It will never be easy but it can be better. Reading book for human relations training* (7th ed.). Arlington, VA: National Training Laboratories.

Royse, D., Dhooper, S. S., & Rompf, E. L. (2003). *Field instruction: A guide for social work students* (4th ed.). New York: Longman.

Schon, D. A. (1983). *The reflective practitioner: How professionals think in action.* New York: Basic Books.

Skovholt, T. M., & Ronnestad, M. H. (1995). *The evolving professional self: Stages and themes in therapist and counselor development.* New York: Wiley.

Sparr, L. F., Gordon, G. H., Hickham, D. H., & Girard, D. E. (1988). The doctor-patient relationship during medical internship: The evolution of dissatisfaction. *Social Science Medicine, 26(11),* 1095–1101.

Stanton, T. K. (1981). Discovering the ecology of human organizations. In L. Borzak (Ed.), Field study: A source book for experiential learning (pp. 208–225). Beverly Hills, CA: Sage Publications.

Suelzle, M., & Borzak, L. (1981). Stages of fieldwork. In L. Borzak (Ed.), *Field study: A source book for experiential learning* (pp. 136–150). Beverly Hills, CA: Sage Publications.

Sweitzer, H. F., & King, M. A. (1994). Stages of an internship: An organizing framework. *Human Service Education, 14*(1), 25–38.

Wagner, J. (1981). Field study as a state of mind. In L. Borzak (Ed.), *Field study: A source book for experiential learning* (pp. 18–49). Beverly Hills, CA: Sage Publications.

Watzlawick, P., Weaklund, J. H., & Fisch, F. (1974). *Change: Principles of problem formation and problem resolution.* New York: Norton.

Wilson, S. J. (1981). *Field instruction: Techniques for supervisors.* New York: Free Press.

Yuen, H. K. (1990). Fieldwork students under stress. *American Journal of Occupational Therapy, 44*(1), 80–81.

CHAPTER *11*

Breaking Through Barriers: The Confrontation Stage

> *If students don't have a chance to address their fears by having the chance to prove themselves, then at graduation they will walk away with the same fears instead of having overcome them.*
>
> *I had a breakthrough happen in what I really want to accomplish in the internship. . . . I can now focus . . . to find some solutions to some long standing problems. Things are moving again and without so much struggle.*
>
> STUDENT REFLECTIONS

After working your way through the exercises in the previous chapter, you probably have a clearer and more concise understanding of the issues, both large and small, that you are facing in your internship. If you think about it, though, you will realize that you have already managed to prevent a number of difficulties from turning into problems. We encourage you to focus on those issues that are still unresolved, without losing sight of your previous victories—however small they may seem to you now—as you prepare to confront those difficulties that await you. This is the Confrontation stage, an opportunity for you to resolve some troubling issues, gain confidence in yourself, and continue to grow.

THE TASKS YOU FACE

The Confrontation stage can be the most challenging time in your placement. It demands that you take charge of your internship in ways that you may not have done before. The process of resolving the issues that continue to challenge you tends to take center stage. As you struggle to achieve a sense of independence, confidence, and effectiveness in your work (Lacoursiere, 1980), you are faced with the reality that such qualities are born not just of your skills and accomplishments but of your ability to overcome obstacles. Breaking through the barriers, for now, is as important as the work in your internship.

Our work has informed us of the trials and tribulations interns undergo while trying to manage these barriers and bring their field experience to a place where they are thriving in the experience rather than just surviving it. To guide you through these barriers, we have developed a model for approaching those problems you choose to take on. It will combine easily with any other models and approaches you may know. Regardless of the approach you choose, though, a model itself is not going to be enough to get you through.

Change can be very difficult. Perhaps you have had this experience either in or outside of your internship. You identify a problem and think you know clearly what is involved. You really want to do something about it. But every time you try to think clearly, you become confused and feel drained, until you finally start to think about something else just to get away from those feelings. Or you may find that you begin making a little progress, only to drown in the frustration and anger that you feel about the problem and your perceived inability to do something about it. This sort of emotional paralysis can be extremely frustrating.

You may feel like you don't know what to do, but knowing is not the problem. And that is the limitation of even the best models. They are largely cognitive; they involve your thoughts more than your feelings. We have found that no model is going to work without three important ingredients—belief, will, and effort—which, interestingly, are driven not by your thoughts, but rather by your heart. The lessons from the heart inform us that the affective domain must be an equal partner with the cognitive domain if you are to be successful in your efforts.

The Power of Belief

If you do not believe that a situation can change, most likely it will not. Let's rephrase that. If you wait for a situation to change on its own, it might. However, the point of this chapter is to help you be an active agent of change, as this intern was: "I have learned that, after you have worked to put all the pieces into place ... what you really want to accomplish finds its way into existence. I just knew it (would happen)."

If you do not believe that you can change a situation, then you probably cannot. Your perceived ineffectiveness will became a self-fulfilling prophecy. Gerald and Marianne Corey have written a marvelous book on self-change and effectiveness entitled *I Never Knew I Had a Choice* (Corey & Corey, 2001). Perhaps *I Never Believed I Had a Choice* more accurately reflects your experience. You need to *believe* that things can be different and that you can create the change that needs to take place. "I had an incident where I didn't agree with the director's decision about a student.... I stood up for what I believed in and what I believed should happen." This intern believed in the potential of the student. It was the intern's commitment to his own beliefs—which were coming from an intuitive feeling about what this student was capable of—that resulted in the student's being allowed to take the upcoming GED. The student did and passed the examination!

Of course, you can work on feeling confident, and that will help you succeed, or, you can work on solving the problem, and that will help you feel more confident. For some people in some situations, it is the belief that is the foundation for the action. For other people, the converse is true. You know best what works for you. However, if you begin to try to effect change and feel your confidence falter, you may want to step back and examine some of the sources of your lack of confidence (a review of Chapter 3 may help you here).

The Power of Will

Will is the intention and determination to make change occur. Intention and determination are powerful motivators, and the frustration and anger you may feel are normal parts of the emotional experience of trying to change something. You need to allow yourself to feel those feelings and not run from them or hide them. As with most feelings, that is how you begin to work through them and allow determination to come to the fore. As one intern recalled, "I . . . just pondered . . . and just wondered whether I was actually helping the client or hindering the client. I was determined to know so I stayed with my decision of saying 'no' and having the client stick it out. She did and I am happy that I did not give in. It would have been so much easier for me to say yes. But, it would not have helped either the client or me."

Another barrier to this determination is the resentment some interns feel about having to work on creating the change. You may feel that this problem is not of your making, and you may be right. So if you didn't "do it," then why should it be you who has to fix it? Accepting responsibility can feel like you are accepting blame for the situation. But you are not accepting blame; you are empowering yourself, as is evident in this intern's comment: "I need to speak up if I am being given too much work. Otherwise I will burn out. Once I start saying 'hold off, I have too much to do.' I know it will affect my internship (in a negative way). But I have to do it." The overriding goal is for you to empower yourself so that you feel able to take charge of your life and your situation when you want to, as this student learned to do.

As long as you wait for others to act differently, you are not in charge. "I learned that I have to take responsibility for my internship . . . to take the bull by the horns as you said. . . . I feel that it's an important step in the process. . . . I did not ask for an agency in crisis for an internship, but whether it is or not, I have got to get my internship on track."

The Power of Effort

Finally, you have to do something. Effort refers to what you actually *do*. It also refers to persistence and perseverance. Some of your difficulties will be relatively easy to analyze and resolve. Others take more time. "What my heart really wanted to do (in this internship) is now going to happen. . . . I am so excited about this great boost (to morale). I feel so motivated now and thrilled to be able to share from my heart something I

Belief, Will, and Effort: A Case Example

Belief, will, and effort, of course, are often interrelated. As you read the following account of a minority student's efforts to bring to the forefront an incident—and maybe a pattern—of questionable practices (from the intern's perspective) in one criminal justice agency, you can see how interconnected these three domains are and how the heart drives the commitment to one's beliefs, will, and efforts.

> *I did an intake with a defendant who was in on a default warrant out of this court from 11 years earlier, which she thought was taken care of because she had been on probation in this same court 4 years earlier and the issue of the warrant never came up. The warrant was not noted on her record and doesn't even come up in the record system. She was arrested on the default warrant after a minor incident over the weekend and held for court after the police ran a check on her. This made no sense, so I decided to make sure it got resolved. Each department I went to said the same thing—let the other department handle it. Did someone forget that the court made this major error and discovered it 11 years later? Now everyone wants to pass the problem onto another department and no one wants to take care of it. She had been a model probationer and never tried to run from anything. I wanted to find out how such a thing could happen and not be noticed until 11 years later. I heard about these kinds of things in my community, and now I was seeing it happen to someone from my community. While I was working in one of the departments with a clerk and checking on this woman's file like I was asked to do by this clerk, the director of the other department came in and asked me why I was involved. I was without words. This was my case. I was checking out her file at the clerk's request. Then I was told by this director to leave the file and leave that office and let that office take care of it. I was bothered. It seemed as if no one cared that this woman and her family were suffering because of the mistake this court made. The good thing is that they all knew that someone—me—was watching and that these matters were not going to be brushed under the rug as long as I worked there."*

This intern's efforts to bring about change actually did make a difference, though the internship ended before the intern got to see those results.

---

know people will enjoy and benefit from. It was a lot of work but worth the effort." Another intern said it more succinctly: "At this point in my internship, it is reassuring to see that persistence does pay off."

By now, some of you may be feeling impatient. "Enough," you may be thinking. "I am ready. I am willing. Let's go!" It will not take long to read the model that follows, but it takes focus and practice to use it well. We hope you are willing to make the effort and that you find the effort worthwhile. But if you are struggling with it and you are sure you understand it, you may want to come back to this part of the chapter and consider these issues again.

A METAMODEL FOR BREAKING
THROUGH BARRIERS:
8 STEPS TO CREATING CHANGE

The mere formulation of a problem is often more important than the solution.
ALBERT EINSTEIN

The literature is replete with techniques for solving various kinds of problems and with generic problem-solving models (see, e.g., McClam & Woodside, 1994). Most models include steps that answer the following questions:

1. What is the problem?
2. What do I know about the problem so far?
3. What is my goal?
4. What are my alternatives for reaching that goal?
5. What is my plan of action?

This generic framework is hardly a new idea; in fact, it has its roots in the ancient teachings of Buddha, who wrote about the Noble Truths, including suffering, the origins of suffering, and the path leading to cessation of suffering (Watzlawick, Weaklund, & Fisch, 1974). Although our model[1] is hardly as profound as its prototype, it does include these basic steps, with a couple of twists to traditional problem solving.

First, the model is designed to slow you down a bit. Yes, you read that correctly. If you have a problem that is really troubling you, you are probably focused on what to do about it, what action to take. Anything, you figure, has got to be better than this. Or you choose a strategy because it worked in another situation or because a friend used it successfully. Thinking intelligently about a problem involves some careful analysis, some goal setting, and finally, deciding on and implementing a plan of action.

This model is also designed to broaden the way you think about your problem. We want you to move beyond the initial stage, where you hold one individual responsible, and look at the perspectives and contributions of many other people. We also want you to consider the wider context in which the problem is occurring, including the systems that are involved, such as work groups, the office as a whole, and other agencies with whom you work. This model asks you to think and feel your way through a process, not simply to plug in answers to questions and in turn get a score that instructs you what to

[1]This model is derived from a model for interdisciplinary integration and problem solving (see Sweitzer, 1989) that in turn has its roots in the Behavior-Person-Environment (BPE) principle advanced by Kurt Lewin (1954).

do next. It also involves a process that you may want to engage in with others. For example, you may want to work with other interns at your site, or you may want to discuss your issue(s) in seminar class using the model for change.

We suggest that you first read through the model in its entirety and follow the example we have included. Next, return to the first step, and, using an issue that is problematic to your internship, work your way through all eight steps. To help you through the steps, we have provided guiding questions and identified potential pitfalls. After you complete the tasks of each step, you will be asked to write a brief paragraph responding to the guiding questions.

ONE STEP AT A TIME

In the middle of difficulty lies opportunity.
ALBERT EINSTEIN

Step 1: Say It Out Loud

Quick—in one sentence—what is the problem? Don't think too much; in the next step, you name the problem more carefully. Just say it and then write down what you said. You can learn something important about how you are thinking from the way you first state the problem. Consider an intern who is struggling with a client. Here are some ways the intern might initially describe it:

- I can't get anywhere with her.
- She is not interested in changing.
- This client needs a different program.
- No one told me how to handle this.
- Their approach isn't working with this client.

Do you see how each of these statements locates the problem in a different place? Where did you initially locate your problem? Where are some other places it might be located?

Step 2: Name the Problem

Now it is time to let go of the feelings and the impulses—the name calling and blaming—and think clearly about the behaviors that are causing problems for you. Don't think about the people; think about the specific *behaviors*. Behaviors are actions that anyone can see if he or she is able to observe. You also need to think about the thoughts and feelings you are having as a result of these behaviors.

Here is one way to do that using a process that will be somewhat familiar to you if you have been using the "three-column" method in your journal entries (see Chapter 1 for review of this method). Take a piece of paper and divide it into three columns. In one, list the behaviors. In the next column, list your thoughts that result from those be-haviors. Finally, list the feelings that result from each behavior. Now, look at what you have written and try to state the problem in one sentence. It can be a long sentence, but try to use just one. Think about when and in response to what this problem usually occurs, and see if you can express that in a sentence. If it is not a recurring problem, this sentence will not be important.

In our experience, though, many of the problems that interns choose to confront in this stage are issues that have come up again and again, although they may appear dif-ferent because the specific problems they create can vary. For example, an intern who has issues with acceptance may choose to eat lunch alone and thus not be included in the energy and discussion that go on in agencies where working lunches are part of the normal routine. Problems can develop if the staff interprets the intern's choice to eat alone as an indicator of a lack of investment or interest in the issues discussed at lunchtime. In the next sentence, see if you can describe exactly how this problem af-fects you. In what specific way does it create unpleasantness? Does it embarrass you? Does it make you angry? Does it hamper your ability to work? Once you have worked on this group of sentences, you will have a concise, clear statement of the problem. These sentences comprise a paragraph. Add it to your worksheet. Here is an example:

> *When she runs away from the program, it angers me because I think about all the work I have done with her, and she just blows it every time. It's just like when I used to work with kids on the streets. They would do the same thing, not listen to me. It makes me feel helpless, sort of powerless, and useless in my work.*

Step 3: Expand Your Thinking

This step is not unlike the reflective observation step of Kolb's learning cycle (Kolb, 1984). You are asked to think about the problem by examining its components from dif-ferent perspectives and along different dimensions. Although taking perspective and ex-amining it are insightful experiences, they can also be frustrating for reasons we have already discussed. But someone has to do the investigative legwork, and in this case, the only one who can do it is you. Begin with some scrap paper. As you work your way through the following four guiding questions, try to keep in mind the role of politics and other subtle dynamics. Be patient and persistent with yourself; some of these steps may not come easily to you, but they all can be mastered with a little perseverance.

1. Who are the players involved in this issue? Identify as many people as possible who are related to the problem. Including them does not mean that they are causing or contributing to the problem, only that they are part of the scene. Don't forget to in-clude yourself.

2. How does each of them see the issue? Work your way through each person on your list and try to put yourself in that person's shoes. Try to see the situation from each of their points of view by imagining what it is like to have their positions on the staff, to have been there as long as they have, to work with an intern, or to have you in their work life.

Some of our students have found it helpful to imagine all these individuals forming a circle, which in turn forms the hub of a wheel. Each person on the wheel sees the issue differently, depending on where they are located. Their perspective, in turn, influences their stand on the issue at hand. Try to imagine how each of these individuals views and feels about the problem. From your perspective, how might each of them be contributing to the problem? How about from their perspectives?

3. What are the major systems involved in the situation?   Each of these individuals is part of and influenced by a number of systems. (You may want to review Chapter 8 if the notion of systems is not yet clear to you.) Your first task here is to make a list of all the systems that are somehow connected to the problem. You may find it easier to approach this task by brainstorming. List all the possible systems you can think of. Begin with those directly connected to the situation and move to those with indirect connections.

Or you might find it easier to be more organized and work your way from the inside out or the outside in. The inside group of systems includes the staff (and any subgroups of the staff you can identify), your work group, the clients, and anyone else who connects directly with the agency. Now think about the agencies that work with yours, either collaboratively, competitively, or adversarially. Next identify the sources of funding and other resources. Don't forget systems on campus, including your professors, the administration, and the seminar class. There may be more; if so, keep going! Now go back to your list and eliminate any systems that, on further consideration, seem to have very little to do with the problem.

Now think about how each system—as opposed to the individuals in it—contributes to the problem. To do this, you need to focus on the rules and roles that drive individual behavior. Again, you may want to review this part of Chapter 8.

4. What about you?   Yes, you! In all likelihood, you are contributing to the problem in some way. Once again, we urge you to return to an earlier part of the book, this time to Chapters 3 and 4 and the exercises you did. Are some of your personal issues and values being touched by this situation? Your initial statement of the problem (Step 1) may be a clue here. Does your response to this situation remind you of any other situation or group of situations? Is there a pattern or a set of irrational beliefs at work?

Now go back over all the ideas and insights you have had during this portion of the problem analysis. Write a paragraph discussing them. Notice how you have begun to create a new picture of the problem that is more complex and takes into consideration multiple perspectives on the issue.

Step 4: Consider the Causes

Go back over your notes and summary from the last step and try to come up with a list of the major causes of the problem. Notice we did not say *the* major cause; there is always more than one. Be sure to consider all the individuals and systems you examined before concluding your list.

The Eight Steps to Creating Change
1. Say It Out Loud
2. Name the Problem
3. Expand Your Thinking
4. Consider the Causes
5. Focus Your Attention
6. Determine Your Goals
7. Identify the Strategies
8. Create the Change

Step 5: Focus Your Attention

You have identified a number of dimensions of the problem. You will need to make changes and alter the dynamics discussed in the cause statement, but where do you begin?

First, try dividing a piece of paper in half and labeling the columns "What I can change" and "What I cannot change." We are asking you to think about the issues over which you do and do not have influence. Let's look at these lists more carefully, this time in terms of what you can and cannot do about the issues. When you do have control over the issue, the responsibility is to do something about it; that is, do what you can do about it. When you have no control over an issue, your responsibility is to accept it; that is, resign yourself to accepting this as something that you must live with.

The possible pitfall in this commonly used exercise is the tendency to confuse what you *can* do something about with what you *cannot* do anything about. As you look over the lists, check for those issues that you continually try to do something about but actually have no control over. Being caught in this danger zone is like spinning the wheels of your car on ice: You are going nowhere fast and using up a lot of energy trying to get there! There are also those issues that you have accepted but can actually have some control over and do something about.

It is time to go back and add a new paragraph to the statement you are developing. This paragraph needs to answer the questions, What is blocking me from resolving this problem? What is it that I need to change?

Step 6: Determine Your Goals

For each cause of the problem on the "changeable" side of your list, you now need to develop goals for change. State as specifically and behaviorally as you can what you want to be different. There is no room for fuzzy thinking here. Just wanting the issue to go away may be your most salient emotion, but now is the time for clear and concrete statements. Be sure that you have covered all the individuals and systems that you believe have a substantial role in causing the problem.

One possible pitfall here is to confuse a goal with an action step. For example, suppose that one major cause you identified is that your duties are radically different from what you expected and were told they would be. Changing placements is not a goal; it is an action step. The goal would be to bring your daily actions more in line with your stated learning goals. One reason this distinction is important is that there are usually multiple ways of achieving any goal. If one doesn't work, take some time to be disappointed, but then think of another. In the example just given, if you made switching placements your action step and then were told you could not make the change, you might feel the problem is now unsolvable. If you keep your real goal in mind, though, you are more likely to come up with another approach.

Step 7: Identify the Strategies

Your task here is to develop a list of possible concrete actions you could take to meet each of the goals you identified in Step 6. In developing the action steps for each goal, be sure the interventions reflect your knowledge about the nature, cause, and context of the problem. The principle of effectiveness is operating here; that is, each action must be effective for each goal. In other words, the action steps you choose must allow the goal to be realized. The picture you have created of the problem in this step is one that is action packed and future focused. When you write your paragraph about this step, let yourself be guided by the question, What must I do to make my goals happen?

You may remember that we called this model a metamodel because it allows you to incorporate many other approaches and strategies. You have probably studied many approaches to resolving interpersonal problems and conflicts, making constructive changes in yourself, and making change at the system or community level. But if you haven't, or if you want to do some more exploration before deciding on a strategy, we have listed some resources at the end of the chapter.

Step 8: Create the Change

This is the step you have been waiting for! Ever since the issue became a barrier to your internship, you have thought about this step. It is the time when change happens. If you've done your work with determination and effort, with genuineness and a willingness to be reflective and introspective in your thinking, and if you asked yourself the tough questions that needed to be asked, then you are ready to move on. It is time to do the achievable and implement your plan with commitment and perseverance. Yes, you've given it your best, and your best is good enough. The potential pitfall of this step is to fail to do what you can do about the issue, resigning yourself to the status quo when you have the capacity to make change occur.

The guiding question for this step is, How will I know when things are different? Let this question guide you as you implement your plan and move through the barrier that has prevented you from moving ahead in your internship. Let it also guide you as you write the last paragraph of your problem statement. Congratulations. You did it!

Not Being Taken Seriously: The Case of the Rebuffed Intern

Some of you may be comfortable with this model and can work with it right away. But for others, it may be too abstract. So we are going to work through an example for you. We will not detail all aspects of the steps, but we will show you how the eight steps work in this particular case.

Leslie is an administrative intern working at a group home that is undergoing a major licensing review. The site supervisor is an administrator. One of the tasks of Leslie's internship is to work with the supervisor in preparation for the review, a task that seemed particularly exciting. However, several weeks into the internship, Leslie was very unhappy and ready to switch internships. Here is the problem as this intern worked through it, written in the style of a series of journal entries.

Step 1: Say It Out Loud

Once able to talk about the problem, I said, "They don't take me seriously!" Just saying the words brought to the surface feelings that had been bubbling below for quite a while.

Step 2: Name the Problem

I was told I would have a major role in preparing for the visit. My supervisor said he would help me understand the regulations, let me do some of the research and even some of the writing, and meet with me regularly to discuss my work. Really, none of this has happened. There is always an excuse and a promise to make it up to me later. I am bored because I don't have anything to do that is worthwhile and makes me think; and I am frustrated because there is important work going on around me and I am not included in it. I'm beginning to think they don't trust me. I spend a lot of time trying to think of what I could have done wrong. I just don't know.

Step 3: Expand Your Thinking

My supervisor
I guess he is under tremendous stress about this on-site review. I suppose he feels responsible for how things turn out for the agency. He is pressed for time. Bet he feels it may be faster to do it himself. I bet he is afraid of his boss. Interesting.

My coworkers
For whatever reason, they don't talk with me about the visit. I see that they are very busy. I know that they had bad experiences with the last intern. But they don't know my skills. Guess they haven't had a chance to see them.

The entire staff
I know that there have been threatened staff cuts. I think somehow that the possibility of cuts creates a "rule" that everyone must make themselves seem as vital and indispensable as possible. That's all that I see—everyone acting important. Must have something to do with the possibility of cuts.

Not Being Taken Seriously: The Case of the Rebuffed Intern *(continued)*

The facility

I know that this review visit is very important to the facility's future. Exactly how, I am not yet sure. I've heard the staff talking about the possibility of declining referrals if there is a problem with the visit. That means their jobs, I guess.

My issues

I know that I am very sensitive to being put off. I really have problems when that happens. I know that it is my issue because other students in seminar class have had similar situations, but they did not make issues out of them. I remember hearing them talking about their situations. I cringed, but they did not.

Another thing I realize is that it is hard for me to be persistent. I actually wonder if I can do this, and I'm terrified I might make a mistake. Just the idea of having to continue to deal with this scares me. You know, I do fine when I have lots of support and attention, but I wilt quickly when I feel the least bit slighted. It really doesn't take much.

Step 4: Consider the Causes

1. I don't have enough faith in myself—I just don't believe in myself at times.
2. I give up too easily!
3. My supervisor is not living up to our agreement.
4. Staff members don't understand my role or what I am able to do.

Step 5: Focus Your Attention

I can't change the pressure that the site visit is causing on my supervisor and on the rest of the staff. I cannot change that this is my internship site (even though I want to at times, but I guess that the field coordinator on campus and my campus supervisor would say, "No—you have got to work this out!"). However, I can change all the causes listed in Step 4.

Step 6: Determine Your Goals

I can think of three primary things that have to happen in order to make this internship work for me. I must:

1. *Increase my assertiveness and self-confidence.* If I do this, I will feel good about myself even if I am rebuffed by others, especially those whose opinions matter a lot to me. After all, not everyone cringes when they feel rebuffed.
2. *Get my supervisor to honor our agreement.* It cannot be that difficult. After all, it is important to him to do a good job, and he does take pride in having an intern. If I can get him to realize that an agreement is an agreement, and that both I and my faculty supervisor believe that I can do the work, things will change for the better. Then I will have something worthwhile to do.

continued

Not Being Taken Seriously: The Case of the Rebuffed Intern *(continued)*

3. *Make sure the staff knows why I am here and what I can do.* I have got to become more visible and better known in the office. Once the staff starts to get to know me, they will want to know why I am in their work space. If I can do that, I will have succeeded in getting them to know me and my work.

Step 7: Identify the Strategies

1. I can learn a lot more about myself if I:
   - Use the trumpet model to work on being able to persist.
   - Use Ellis's ABC model to understand why I wilt so quickly.
2. I can get my supervisor to honor our agreement and the contract if I:
   - Request a special meeting.
   - Write my concerns down ahead of time.
   - Get support from my faculty supervisor.
3. I can make sure the staff gets to know why I am here and what I can do if I:
   - Ask my supervisor to invite me to a meeting and introduce me.
   - Volunteer to help the staff with some of their regular duties.

Step 8: Create the Change

Believing I can bring about change, I must persevere and continue until I see the changes that suggest:

1. I am feeling more aware of my role in bringing about change and more confident about my work and the future of my internship.
2. My supervisor is more aware of how the internship is affecting me and is willing to reconsider the learning contract we created in order to better meet my needs.
3. Staff members are more invested in their relationships with me and are actively reaching out to involve me more in their work.

I guess I am surprised at what I can do to make my internship different. It really is not all that difficult. Makes me wonder why I have been stuck for so long.

SUMMARY

This chapter has prepared you to confront the issues that challenge you and has given you some tools by which to do it. However, sometimes it is also helpful to pay attention to how others approach and resolve problems. Look around you. You probably won't have to look far until you find someone—a friend or a peer—who thinks very differently than you do about problems and their solutions. What are those differences? How might others have dealt with the issue(s) you are facing? Can you learn anything from their approaches?

We close this chapter with this student's thoughts:

I never thought the crisis I experienced and the change I brought about could be used as a selling point on my cover letter. . . . The whole thing has made me realize how beneficial this crisis has been for me down the road.
STUDENT REFLECTION

For Further Exploration

ON PROBLEM SOLVING

McClam, T., & Woodside, M. (1994). *Problem solving in the helping professions.* Pacific Grove, CA: Brooks/Cole.

Clearly written and concise with a problem-solving model and lots of cases to study.

ON INTERPERSONAL EFFECTIVENESS

Corey, G., & Corey, M. (2001). *I never knew I had a choice* (7th ed.). Pacific Grove, CA: Brooks/Cole.

Excellent book with a good chapter on relationships.

Johnson, D. W. (1999). *Reaching out: Interpersonal effectiveness and self-actualization* (7th ed.). Upper Saddle River, NJ: Prentice Hall.

Clear and incisive with lots of examples and exercises—a classic.

ON MAKING INTRAPERSONAL CHANGE

Brill, N., & Levine, J. (2001). *Working with people: The helping process* (7th ed.). New York: Longman.

Self-understanding and self-modification are discussed throughout this book.

Ellis, A., & Harper, R. A. (1975). *A new guide to rational living.* Upper Saddle River, NJ: Prentice Hall.

Ellis's ABC model is explained with helpful self-help guides to work through your own situations.

Gibbs, L., & Grambrill, E. (1996). *Critical thinking for social workers: A workbook.* Thousand Oaks, CA: Pine Forge Press.

Provides a comprehensive skills-based approach to develop reasoning skills for effective decision making.

Grobman, L. M. (Ed.) (2002). *The field placement survival guide.* Harrisburg, PA: White Hat Communications.

Offers social workers an edited volume of writings that include skills and advice in confronting challenges.

Sweitzer, H. F. (1993). Using psychosocial and cognitive behavioral theories to promote self understanding: A beginning framework. *Journal of Counseling and Human Service Professions*, 7(1), 8–18.

Combines Erikson's work with personal reaction patterns and discusses implications for human service work.

Weinstein, G. (1981). Self-science education. In J. Fried (Ed.), *New directions for student services: Education for student development* (pp. 73–78). San Francisco: Jossey-Bass.

An easy-to-follow model for interrupting reaction patterns that may be troubling you.

ON SYSTEMS CHANGE

Homan, M. (In press). *Promoting community change* (3rd ed.). Pacific Grove, CA: Brooks/Cole.

Excellent, well-written text on community organizing with interesting thoughts on the change process.

Horejsi, C., & Garthwait, C. (2002). *The social work practicum: A guide and workbook for students.* Needham Heights, MA: Allyn & Bacon.

Offers a comprehensive approach across contexts of practice for skills development in bringing about change.

Kotter, J. P., & Cohen, D. S. (2002). *The heart of change: Real-life stories of how people change their organizations.* Cambridge: Harvard Business School Press.

Offers an approach that affirms the importance of teamwork, visions, and communication in bringing about short-term wins that create change.

Shepard, H. A. (1975/November). Rules of thumb for change agents. *Organization Development Practitioner*, pp. 1–5.

Offers aphorisms to think about when you are the change agent.

Watzlawick, P., Weaklund, J. H., & Fisch, R. (1974). *Change: Principles of problem formation and problem resolution.* New York: Norton.

Somewhat theoretical but very interesting. A groundbreaking book on why change can be so stubborn and what to do about it.

References

Corey, G., & Corey, M. (2001). *I never knew I had a choice* (7th ed.). Pacific Grove, CA: Brooks/Cole.

Kolb, D. A. (1984). *Experiential learning: Experience as the source of learning and development.* Upper Saddle River, NJ: Prentice Hall.

Lacoursiere, R. (1980). *The life cycle of groups: Group developmental stage theory.* New York: Human Sciences Press.

Lewin, K. (1954). Behavior and development as a function of the total situation. In L. Charmichael (Ed.), *Manual of child psychology* (2nd ed.). New York: Wiley.

McClam, T., & Woodside, M. (1994). *Problem solving in the helping professions.* Pacific Grove, CA: Brooks/Cole.

Sweitzer, H. F. (1989). The BPE framework: A tool for analysis and interdisciplinary integration in human service education. *Human Service Education*, 9(1), 11–19.

Watzlawick, P., Weaklund, J. H., & Fisch, R. (1974). *Change: Principles of problem formation and problem resolution.* New York: Norton.

GOING
THE DISTANCE

You have come quite a distance in your internship. You have dealt with the concerns of beginning your internship, you have launched and nurtured a number of relationships, and you have learned how to identify and cope with problems. The next several weeks ought to be a time of great productivity and enjoyment, although they will undoubtedly have their challenges. Chapter 12 examines in more detail the joys and challenges of the Competence stage. Along with this exhilaration, reality is ever present as you begin to take notice of the complexity of the professional and political demands of the work and the work place. Chapter 13 provides you with an overview of the types of ethical and legal concerns that you begin to encounter now that you are grounded firmly in the work. Another aspect of reality that creeps into consciousness at this time are thoughts of ending the field placement; this is

yet another critical time in the internship. End-
ing well is not easy, but it can be done, and in
many different ways. Chapter 14 will help you
meet the concerns and challenges of the final
stage, that of Culmination.

CHAPTER 12

Riding High:
The Competence Stage

Things that didn't seem to come together have finally come together.
STUDENT REFLECTION

The price of greatness is responsibility.
WINSTON CHURCHILL

There comes a time in almost every internship when interns tell us that they are enjoying their internships in a whole new way. The initial anxieties have subsided, problems big and small have been resolved, and considerable confidence has been gained. There is a sense of feeling grounded . . . finally. If this sounds like what you are experiencing, then you have entered the fourth stage in internship, the Competence stage. Like any stage, this one has its concerns and its challenges, but for many interns and supervisors alike, it is the most exciting and rewarding time in the field experience.

Students have described this stage as the *emotional moment* of the internship. It is this stage—the Competence stage—that defines the internship for the students. Finally, they *know* what they are doing. The interns tend to work independently at this point and make decisions as needed, under supervision, of course. The learning contract is being fulfilled during this stage. The interns feel in charge of their work, and they feel in charge of their internships. Marked by "good" feelings, this is the stage when the interns feel confident about their knowledge, their skills, and their competence.

We use the concept of *competence* to refer to the intern's overall productivity and achievement, including interpersonal and intellectual skills (Chickering, 1969). The intern's concerns during this stage focus on developing a sense of competence and taking charge of the internship to ensure a quality experience. Striving for competence often means being all that you can be in your role as intern and doing your very best. You may recall the discussion of competence in Chapter 3, and Erikson's (1963) position

211

that a sense of competence includes feeling like a capable person even when you stumble or fail. Most likely, you have experienced times during your internship when you were disappointed by your performance in some way. According to Erikson, if you have a strong sense of competence, moments like these will not shake the basic confidence you have in your ability to succeed and move ahead.

You may realize by now that you have grown not only in terms of your knowledge and skills but in personal ways as well. The transformation that is occurring may seem slow and subtle, so much so that you didn't notice until someone commented on it. Or, it may seem sudden and dramatic, leaving you to wonder what is happening.

Either way, you are changing. You have successfully faced new situations, including ones that seemed overwhelming at the time; you have accepted greater and greater responsibilities, developed your competencies, and enhanced your faith in yourself; and you have learned to develop new ways of thinking about situations and about the people with whom you work, including yourself. None of these changes has come easily, but they add up to a transformation that lets you see more, do more with what you see, and be more in charge of your internship. In short, you are empowering yourself, and it shows every day in your work and your approach to it.

Students arrive at this stage at different times in their internships. Some of your peers may not be here yet, while others may have been here for some time. What is important is not how fast you reach this juncture, but rather having the confidence in yourself that you will get here in due time. As you join your peers in reaching the Competence stage, keep in mind that you have different styles of expressing your emotions (see Chapter 10), and while you may experience this stage as a subtle, yet profound, shift in feelings, others may experience it as a positively giddy swing in emotions. In this chapter, we review some of the pleasures you can expect from this stage and prepare you to face the challenges that lay ahead.

ENJOYING THE RIDE

This is the time in field placement when your experience most closely matches the images you had in the beginning of what an internship was all about. You finally are doing what you set out to do. You have greater responsibilities each week (and may even be wanting more!) as you strive to reach your goals, and most of your energy these days is devoted to fine-tuning your skills so that they become second nature to you, and you can generalize them to everyday situations.

Chances are that you are also seeing your internship through a set of new and different lenses. Having adequate knowledge to feel competent in doing the work is critical during the Competence stage. It's not as simple as what you know and what you don't know. Discovering knowledge you didn't know you had and discovering gaps in knowledge you thought you had can be overwhelming at times. One intern described it this way: "I went to a training session yesterday . . . that made me think about my approach with all of my clients and how far I have to go. I found myself thinking about this new knowledge and how learning this little bit of knowledge was just the tip of the iceberg. It was all the stuff I didn't even know I didn't know!"

This is often a time when interns are able to differentiate more clearly between what is important to their work and what is not (Dreyfus & Dreyfus, 1980). Subtle dynamics that escaped you before are probably visible now. Remember when you were concerned about acceptance issues? "Now I look back on all those feelings of inadequacy and laugh!" was one intern's reaction to that period in her placement. Even though you spent quite a bit of time trying to find your way around the organization when you were concerned about these issues, the organization's politics were virtually invisible to you because your primary concerns were focused elsewhere. Now, however, you are more savvy and in tune with the ways of the organization, and you feel the effects of office politics when they are affecting your work. Why? Because your primary concerns now are on the quality of work and your competence in doing it. Office politics can affect both of these dimensions.

Your emotional landscape is changing as well. Many interns feel an enhanced sense of mastery as their skills become sharper and their sense of autonomy requires less supervision and direction. There is also a tremendous sense of confidence associated with this stage. The anxiety, awkwardness, and trepidation of earlier stages have gradually given way to a period of calm and inner strength. Interns often tell us that things have settled down at the site. However, life at the agency is still hurtling along at the same pace. It is not the world around you that has changed and become more peaceful; rather, it is you who have developed an inner sense of calm from which you will derive strength as you move through this intensely productive period in your internship.

Redefining Supervisory Relationships

The importance of the supervisory relationship cannot be overstated: It is the aspect of the internship that our students have consistently reported over the years as most critical to their success in the field. We are sometimes struck by the quality of the relationships our students develop with their supervisors. They not only are supportive and encouraging connections, but they often become mutually satisfying relationships that last long past graduation day.

When we ask our interns to reflect on the changes that have occurred in their field placements, they immediately identify changes in relationships with their supervisors. They smile as they tell us that they no longer need to check in with their supervisors as they go about their work. "At the beginning of my internship I was constantly asking, 'What do I do next? What do I do now?' Now I am not asking what to do, I am just doing it." Supervision is more predictable now and tends to occur at weekly, structured, individual or group meetings. This shift in frequency of and dependence on the supervisory meetings can be most affirming for emerging professionals like you.

Working with staff and supervisors becomes easier during this stage. Communication becomes more comfortable and open, issues can be approached without concern about rejection or conflict, genuine teamwork can develop, and supervision can be a source of insight and feedback for personal as well as professional growth. Often, a change in the openness of the supervisory relationship becomes evident at this time as well. Most of you would not have considered challenging your supervisors on theoretical grounds earlier during the placement. However, a combination of confidence and a sense of emerging equality may make a big difference in your willingness to engage the supervisors in academic discussions, as well as in disagreements (Lamb, Baker, Kennings, & Yarris, 1982).

There has also been a shift in your primary focus. Chances are you are talking less about yourself during supervision and in seminar class, and more about the needs of clients or the community. This is just one more shift that tells you important changes are taking place (Tryon, 1996). One consequence of such shifts is that your supervisory needs are changing (Hersey & Blanchard, 1982; Tryon, 1996). Your former dependence on your supervisor for direction and support has given way to working more autonomously and developing more equal relationships with your supervisors (see Chapter 7 for a review of supervision issues).

Some site supervisors develop more than a supervisory relationship with their interns. They develop what is traditionally called a *mentoring relationship*. The mentoring relationship differs from supervision because it is interpersonal in nature and is one in which the supervisor takes an interest in the intern's professional development and career advancement (Collins, 1993). Many studies have been conducted to understand what happens when this type of relationship develops. It can be a very empowering relationship, leaving both the mentor and the mentee or protégé better professionals for it. It may interest you to know that the mentoring relationship is often considered the most important aspect of graduate education (Mozes-Zirkes, 1993). At the undergraduate level, we have found that the mentoring relationship, when it develops, is a most instructive and influential factor in the quality of the internship experience.

The best mentoring relationships occur spontaneously between the supervisors and the interns. The effective mentor makes sure that the intern becomes part of the organization very quickly and is given highly visible tasks; this type of mentor ensures that the intern is introduced to the profession through such resources such as networking, luncheons, and conferences. Although many supervisors make sure that all of these bases are covered for the intern, the mentoring supervisor personally invests considerable time and many resources in coaching the intern for success (Tentoni, 1995). In other words, there is a personal investment on the part of the mentor in making the mentee successful.

Of course, this relationship, like others, takes time to develop (Collins, 1993). It moves from a relationship of positive role modeling, when much of the learning occurs through interactions, observations, and comparisons with the supervisor, through a time when you have grown to really like each other, and then to a point where you might be right now: of valuing the relationship and recognizing how mutually rewarding it is for you and your supervisor. When this happens, the intern tends to become increasingly self-assured, competent, and autonomous, no longer needing the mentor in the same ways for guidance or support, and eventually becoming more independent of the supervisory relationship. In the best of mentoring relationships, a more equal relationship emerges over time. That becomes evident when the supervisor accepts you as a colleague, and you accept the supervisor on equal footing (Collins, 1993).

Redefining Your Identity

A second area of significant change is a most visible one: your career status. A transformation is taking place, and it is a change in your status from that of a student to that of an aspiring professional. You think differently, you act differently, and you feel differently. How did these changes happen?

You may not be aware of this, but how you think about yourself is continually redefined by your experiences in the internship and the situations you face every day at the field site. Have you noticed that you no longer refer to yourself as "just an intern" or "the intern"? Rather, you recognize yourself more and more as staff and as a member of the profession. One intern reported, "I am fitting in around the office. Everyone is helpful and giving me tips on how to do things. They even ask me what I would do in certain situations. I feel they respect what I have to say." Another offered, "I realize that people really do depend on me here. . . . My advice is important." You have a greater sense of self-awareness and self-respect. You are most likely learning that you can trust your knowledge and skills, and you are developing a good sense of your strengths and limitations. Your work might suddenly be taking on new meaning because you realize that you are making a difference in the lives of others. Your professional identity is beginning to emerge.

There are two hallmarks of this shift in identity that are discussed with frequency by our students: a commitment to quality work and a commitment to personal integrity. Not so long ago, in the throes of the transition from learning the work to doing the work, "good enough" may have been all you expected of yourself in your endeavors; just getting the work done seemed a lofty goal! That was especially so if you were dealing with a multitude of responsibilities in addition to your internship, or if you were experiencing the confusion and discouragement associated with the Disillusionment stage. Now, though, the "good enough" standard that prevailed earlier in the internship for some of you gives way to a standard of excellence as you enter the Competence stage. You become more demanding of yourself, regardless of the amount of responsibility you have. This expectation of excellence extends not only to all aspects of your internship, but to those aspects of your personal life that are directly connected to the field experience.

We do not think an internship necessarily changes your basic core of values; however, it certainly does push you to clarify them professionally and personally. When this happens and you are accomplishing a great deal, you may feel so proud of your work that you want to keep on doing it and doing it better. A word of caution here. Many students have learned from experience not to confuse the need to do work well (excellence) with the need to do work perfectly (perfection). Though both may guarantee success, one guarantees headaches as well—for all involved. If perfection is your only or preferred way of doing things, you are bound to have problems.

The second hallmark of a changing identity is a more conscious commitment to a sense of personal integrity. Changes here are reflected in the ways that you think (flexibility and adaptability), what you value, and how your beliefs and values affect your behavior (Chickering, 1969). Although you may feel that you have matured substantially since you began your internship (much to your relief and, in some cases, to that of your loved ones and family!), changes in your basic level of integrity tend to be gradual and do not affect the values that are important in your life. Developing integrity is a lifelong process that begins during the college years (Chickering, 1969). This is not a goal of your internship, but rather a gift that emerges from the maturing process.

As you grow in integrity, your values become genuinely reflected in how you act. In other words, your behaviors truly represent your values. For this harmony or congruence to occur, it means letting go of your literal beliefs in the external rules you live by and re-constructing a new value structure for living. This is no small feat, and in fact, you already are working on it. By doing so, you are "humanizing values," which is essential for ongoing maturity. Though necessary, letting go of one's literal beliefs can be quite perplexing when in the throes of a life transition, such as the internship is for many students.

Because anxiety and tension can run very high at these times, students develop an internal structure to make sense of their world. This new structure that emerges is one that represents the essence of you. It is not your parents', your school's, or your friends' ways of thinking, but yours and yours alone. You own it (emotionally speaking); it is a part of you and it reflects your values and beliefs. The closer you are to adopting this value structure openly and without façade, the closer you are to living with a sense of personal integrity (Chickering, 1969). This is a lifelong task, and your journey has just begun.

BRACING FOR THE BUMPS

Although the Competence stage is recognized primarily for its good feelings and intrinsic rewards, it has its challenges like all other stages in the internship. For some interns, the highs of this stage allow the challenges to be put into perspective and managed rather easily; for others, the challenges seem to threaten the good feelings that characterize this time in the placement. One intern summed it up this way: "This incident helps me put things into perspective…(and) makes me realize more and more that as successful as this internship is and will be, there will be several bumps in the road." These challenges cannot compare to the ones you faced in the past; those felt like walls that stopped you from moving forward, whereas these are more like bumps along the way.

These bumps in the road can throw you off course if you do not know what to expect or how to manage them at the time. We have identified three potential bumpy areas that you need to pay attention to: (a) the tendency to focus on only the good feelings and avoid the challenges that are an inevitable part of your placement; (b) the tendency to not enjoy the success you are achieving; and, (c) the tendency to experience predictable transition issues during this highly productive time in your internship.

Freezing the Moment

Many interns fall prey to the misconception that the days of difficulties or problems are behind them. Some say they expect more difficulties in their internships, but when the difficulties arise, they ignore or fail to acknowledge them. Other interns just want to focus on the good feelings of having resolved issues. And still others fall into the trap of believing that real professionals (like really good interns) should be beyond having difficulties.

Well, you are not beyond them—no one is. Your skills, your relationships with clients, supervisors, peers, and coworkers, and your understanding of the field site are all evolving, and you could continue to experience new difficulties in all these areas.

The good news is that you are better equipped than ever to meet these challenges; in fact, your ability to handle difficulties can be a source of ongoing pride. The difficulties you face at this time may actually feel different now and not be so overwhelming or anxiety provoking. However, they still need to be managed or the difficulties will become problems.

Experiencing Success

Some interns become concerned and challenged at this point in their internships when they realize that they appear successful to others but they do not *feel* successful about their accomplishments. Such a situation can become problematic because your experience of the internship is based on what you feel, not on the perceptions of others. So feeling good about all aspects of the internship is very important to your sense of success. Being able to feel your success, often described as a sense of fulfillment, is what makes the success genuine; anything less can compromise it (King, 1988).

What is particularly interesting is that this sense of inner success, which reflects your success as a person, is believed to render all of your achievements worthy (Huber, 1971; King, 1988). Not only do you need to achieve the goals of your internship (outer success), but you also need to pay attention to yourself as a person (inner success) if you want to reap such psychological rewards as the internship being "the best experience of my life" and "a most fantastic experience." In the next part of the chapter, we discuss three sources that nurture the emotional experience of success (King, 1988). As you read about them, think of your own field placement and the ways it provides you with what you need to feel the success you have earned. Please refer to Table 12.1 that follows on page 219.

SOURCES OF FULFILLMENT

The first source of fulfillment in success concerns the "outer," or social, dimension of success (Huber, 1971) and involves the work you do in your internship every day. Whatever your work might be, it must be considered *worthwhile work* by you and your supervisor. The work itself needs to include productive and responsible activities that have personal meaning, and it must allow you to accomplish clear goals.

Worthwhile Work   There are five aspects of the work that allow these feelings to develop: accomplishments, acknowledgement, self-determination, self-actualization, and intrinsically rewarding work. First, the goals of the learning contract need to reflect *accomplishments* that are purposeful, constructive, and challenging, and which can be a source of pride for you. The work itself needs to provide you with opportunities to develop your skills and apply concepts in real-life situations. For your part, you need to take an active role in creating your work and your achievements. "By allowing me to take charge and work independently with a group of clients yet having the support of a professional . . . I am proud of my internship now that I feel like I'm making a difference and doing some meaningful work" was one student's reflections on her accomplishments.

Values also play an important role in the sense of accomplishment you feel about your work and your association with the agency. Your values and those of the field site need to be compatible for you to feel like a part of the organization. In addition, both your work and that of the agency need to reflect socially responsible goals to have credibility in the community. Interning at a site that is a known abuser of the environment or has a reputation for questionable practices with clients would certainly call into question—by you and by others—the quality of your success.

Second, it is important that you be *recognized* and *respected* by your supervisor; such acknowledgment is especially significant when your contributions are above and beyond what is expected of an intern. The third important aspect of the work has to do with the potential for *self-determination* (autonomy); for example, it is important that you have the freedom to create and carry out tasks. "I coordinated the event and actually made it happen.... I was given the opportunity to develop something on my own. It felt great" was how one student thought about it.

The fourth aspect has to do with the potential for *self-actualization* in your work; for example, there need to be opportunities for creative expression and personal growth in the work you do. And, above all, the work has to be *intrinsically rewarding*, such as when you make a difference in the lives of clients and communities. Then, regardless of the demands or stresses, the sense of intrinsic worth will allow you to derive pleasure from what you do. "It made me feel good to help someone and it was nice to be appreciated for my efforts. I feel like I am making a difference at my internship. It feels good."

Responsible Relationships   The second source of fulfillment in success concerns the "inner," or personal, dimension of success (Huber, 1971) and is derived from the personal relationships that are integral to your experience of the internship. For your relationships to be a source of feelings of inner success, they need to be genuine, cooperative, and mutually satisfying for all involved; in other words, they need to be *responsible relationships* (King, 1988).

Real commitment on your part and on the part of staff, coworkers, peers, and supervisors is needed to make responsible relationships happen. The staff must be receptive to your coming on board, and when it happens, it is an important milestone. This intern speaks to its effects: "This past week I have been looked at as a member of the staff, not just by the residents, but also by the staff. That was so important." The supervisor needs to establish an effective supervisory relationship with you and include you in work groups or teams, staff meetings, and social functions (when appropriate), and you in turn need to be willing to engage both the supervisor and the staff in responsible supervisory and collegial connections.

Self-Defined Success   The third source of feeling the success you attain is being able to *define success for yourself*, which means you are doing what you want to do. This choice must be yours and yours alone, and it must be a conscious choice made free of the influences of other people (King, 1988). When it comes to your internship, it means you are have been actively involved in the selection of your field site and in the development of the goals of the learning contract. If the field instructor determined your site and the site supervisor determined your learning contract, and you were not actively involved in an informed way in these processes, the likelihood of your feeling successful about your internship probably has been compromised.

However, all is not lost here. Even if you went into the internship not consciously aware of what you wanted, those overseeing your placement may have been able to assess your interests and goals well and place you with a supervisor who provides considerable support for personal growth and awareness, while at the same time entrusting you with worthwhile work. If that is the case, then there is a good chance that eventually, you will emotionally *own* the choice of field site and the learning goals, and, in turn, begin to feel successful in your achievements.

You will need to do some soul searching about this, and now might be a good time. Think about what you really wanted to achieve in this internship, not what others think you ought to achieve or what others want for you in this experience. Think as well about the extent to which your placement site is a good match for your personal goals and a site you might have freely chosen for yourself. And, think about the extent to which the choice of internship was made freely by you. These are the thoughts of one student who did engage in such soul searching: "Oh, my…I'm not sure that I even want to be a counselor now! After all this—night school for years—all this work. And, putting off starting a family. But I have to realize that every agency is different. Maybe the way this site is run is different from others. Maybe."

So, what do you do if you are not feeling your success? Actually, there are several things you can do. First, you can spend some time determining whether or not the design of your internship allows you to experience the steps of processing and organizing experiences so that the learning cycle described in Chapter 3 takes place (Kolb, 1984). How does the design of your internship measure up to that framework? Next, you can spend some time examining the relationships with your supervisors and others who are important to your field experience. Are they disappointing relationships in any way? If so, why do you think that is? If not, what makes them so good? Finally, you can review the sources that nurture the feelings of success and, using the "Steps to Create Change" from Chapter 11, identify the areas in your internship that you would like to see strengthened.

TABLE 12.1
Sources of Fulfillment in Success for Interns

Doing Worthwhile Work	Developing Responsible Relationships	Defining Success for Yourself
Accomplishments	Supervisors	Conscious Choices
Acknowledgment	Peers	Active Involvement
Self-Determination	Staff	
Self-Actualization		
Intrinsic Rewards		

Encountering Transition Issues

There are several issues that tend to surface as you enter this period of highly productive work. They all seem to be part of a natural transition to a more realistic way of living with the internship once the crises in growth are over and the climb has ended. If these issues are not recognized and understood for what they are, they can compromise your feelings of competence and the joys that come with this stage in the internship. A discussion follows of three of these issues: (a) the leveling-off effect, (b) the need for a more balanced life, and (c) the inevitable "crunch."

The *leveling-off effect* is experienced as a subtle change in the pace and intensity of your ascent up the ladder of responsibilities. Students often liken it to that point in an air flight when the plane is no longer ascending and begins to level off. The timing of this change in the internship is determined either by the calendar or by a saturation in workload. When this happens, you will experience the cruising effect: The intensity of increasing responsibilities decreases and you no longer feel the constant push to greater heights. Eventually, you will stop taking on new responsibilities and settle into a rhythm of working that is more realistic and predictable in nature. Here's how one intern described his experience: "This was the first week that I can think of that nothing new happened at all. I need to learn that sometimes in this line of work there will be days and weeks where nothing big will happen. So having this happen was a good thing because it's teaching me things that I need to learn, but I don't want to learn."

For many students, leveling off means that the goals of the internship are being reached, and it is time to relax somewhat and enjoy the ride. For some, though, it is difficult to enjoy the ride because of the mixed feelings that come with reaching the final plateau of a journey. For a brief moment, you might wonder whether this is all there is to the internship and speculate that there are few surprises left in the work and little left to learn. Be assured that any emotional letdown you experience in reaching the apex of your journey is only momentary and will pass. You will adjust to these changes and get on with the challenges and personal rewards that come with being a productive member of the staff.

You are probably sensing by now that your total immersion into the role of being an intern is beginning to change (Lamb et al., 1982). If you are like many interns, you may suddenly realize that you had a life before the internship—a social life, a family life, a private life—and you find yourself needing to reclaim that life. One intern was stunned with this realization: "I realized I was not giving enough attention to my relationships with my friends and family. It had been nearly three months since we had spoken. The incident made me stop and think about how important my relationships are and how, even though I am busy, I must not neglect them because these are *the people* I count on for support."

Perhaps that awareness will come to you on your way home from the field site, when you are no longer so preoccupied and are thinking about aspects of your life that were left behind in all the excitement of the internship. Or perhaps it will happen during a morning break as your mind wanders to more personal needs—for relationships on-site or in your personal life, for family time, for time alone. "I can really understand the need now to slow down and pace myself and try to create a more livable, useful pace for other people in my life, including at the site," noted one intern. Another student's experience was a nagging feeling. "I can feel a tug going on between family demands, work, and my academic requirements. . . . The holidays can also complicate my situation to the point where I am beginning to feel both pressure and stress."

Interns usually find a need to get on with living at this point in their internship. In our experience, these changes signal a healthy shift of energies toward a more *balanced way of living* with your professional life. Though your internship still remains very important, it may no longer be the driving energy of your life.

Up to this point in your internship, you have had to struggle with the ups and downs of becoming part of an organization, learning skills, developing competencies, resolving differences between real and ideal expectations, and contending with changing perspectives and life's big questions. Just when you thought it was clear sailing ahead . . . *wham!* You've run into yet another wall, and just about everyone hits this one.

This wall is different, though. You know you can manage it; you just don't know whether you can muster the energy you need to manage it. You feel some anger about the amount of work you have and frustration over the amount of time you do not have. Something subtle is happening as well. Your perspective seems to be changing again. You have little tolerance for trivialities, and you find yourself feeling indifferent about some of your assignments and responsibilities, especially if they are not directly related to the internship or do not measure up to the importance of the work you are doing in the field. This is not a crisis in growth you are having; this is a *crisis in management* of your time and the workload. The crisis takes on added importance because it is in fact a real threat to your schedule and to the pace of your internship.

You are encountering what we refer to endearingly as *the crunch*. It is an early warning signal to let you know that the end of the internship is fast approaching. This is a crisis in management. It really is no different from the crunches you faced at the end of every semester (usually at the end of the mid-semester slump, remember?) when papers, exams, and projects all become due at the same time. However, this time, much more is at stake academically, financially, and professionally. And, you also are carrying the greatest number of responsibilities you have had in your academic career. Have you noticed . . . you have begun to question the merit of each assignment? Challenge all requests for more work? Fear crashing into the wall of professional ineptitude? You are beginning to feel overwhelmed and frozen in time . . .

too many responsibilities . . .

too many deadlines . . .

too many details . . .

too few resources . . .

too little social life . . .

too little family life . . .

too high standards . . .

Oooops! You're beginning to slide . . .

If any or all of the preceding items best describe your internship at this time, you are not alone! Most of your peers are going through the same crunch in their internships.

The time has come to rethink your situation and regroup yourself, emotionally and physically, for the last mile of the journey. You know yourself best in times like this.

What works for you? What allows you to manage time and workload crunches? Some people actually thrive when they are up against a wall of deadlines, while others crumble under the same circumstances. Perhaps one of the following suggestions will work for you, as they have for many of our students.

Take time for yourself   A day will do; an afternoon or evening could do just as well. The important thing here is to *stop*. Just stop everything. Take a holiday from the assignments, the responsibilities, and the timelines you are facing. Of course, it is not just the internship that is overwhelming you. Many of you have many more responsibilities beyond your internship, and you may be overloaded for what you can effectively handle at this time. However, the internship is still the new kid on the block, and its workload is most demanding right now. It is very important that you do something that you enjoy. Sing. Dance. Shop. Bike. Hike. Climb. Swim. Draw. Paint. Read. Run. Play. Ski. Work out. Create. Sew. Meditate. Skate. Listen. Make music. Take in a movie. Go to dinner. Do something, anything, whatever it takes to help you let go, relax, and have fun. Try to do something relaxing at least weekly.

Prioritize and schedule   You must look at the calendar and make some decisions. It is time to prioritize your tasks because things simply must get done. What you will need to determine is what you can do in the time you have and what can get done only if you make big changes in your schedule. Here is an exercise that our students find helpful in accomplishing this task. First, list all the tasks that you have to complete. Then, next to them, indicate the time frame within which they *can* be finished, not necessarily *must* be finished. For example, it you have a final culminating statement to write for your seminar class, it obviously cannot be completed before the internship ends. However, you can work on much of it before the end of your placement. Then, when the internship is over, there is relatively little content to add. The time frame for this assignment actually begins on the date you start this exercise and ends on the due date. By listing tasks, assignments, and responsibilities in a similar manner, you will have a more informed framework for prioritizing tasks and assigning calendar dates to work on them.

What is obvious is that something has to change to put you back on track and in charge of your internship. You already have had to deal with difficult challenges when you confronted the barriers to your internship that prevented you from moving ahead. This hurdle is far less complex and demanding and actually quite manageable by you simply, if for no other reason, because you have gotten this far in your internship. So, if you are struggling with this challenge, roll up your sleeves and get to work. Using the model of Eight Steps to Creating Change (see Chapter 11) will help you clarify the biggest hurdles you face during this time-and-workload crunch and bring this crisis in management to an end.

DEVELOPING
PROFESSIONAL AWARENESS

As the headiness of the Competence stage continues to define your daily feelings about the internship, you may find yourself thinking more and more about the profession for which you are preparing. Preparing your résumé, seeking out job opportunities, and deciding where you want to live after graduation perhaps comes to mind immediately. And, those are certainly important ways to begin your journey to your professional home.

We find that interns are preparing themselves in other ways as well, ways that reflect their growing awareness of the importance their profession will have in their lives. One of those ways is a need to leave their mark by *giving back* at their internship site, a need to feel a part of the organization and profession after the internship ends. It certainly is a way to ensure that you are not forgotten at your site. Whether you are in a public service setting, corporation, civic organization, research laboratory, or nonprofit agency, you may be feeling a need to make contributions to the field site in meaningful ways. As one intern put it, "I think, no, I *know* that the (police) station needs to have the back room cleaned out. I want to do that for the chief. He has been very supportive and I want to give back in some way.... I guess it means a lot to me to know that I somehow am still going to be part of the department in some way after I leave."

These contributions are not part of the learning contract, but tend to be given from the heart. It is a contribution the intern sees benefiting the agency, and because of the workload demands, fiscal constraints, and task priorities, the work would have to wait until some other time to be realized. The contribution may take the shape of creating an internship manual for the site, developing data banks, creating a resource library, or reorganizing workspaces. A popular contribution for our students has been reorganizing and downsizing data files that need to be recataloged into new data systems. For the most part, the contributions tend to be on the "wish lists" of supervisors or agency directors. If you have been thinking about making such a contribution, now is a good time to think about how to do it and when. You might want to use the predictable down time during the last few weeks of the internship.

A second area of awareness that typically develops in students during the Competence stage has to do with an appreciation for what is beyond textbook knowledge. By now, you realize that remembering facts, working with efficiency and speed, approaching problems in logical ways, and effectively using the environment to meet the needs of the work are effective ways of going about your work. In fact, they may be exactly what will be expected of you in your future profession. But you also realize by now that it takes more than these skills to handle challenging situations and resolve complex problems in the workplace. For example, if your field site encourages and affirms a tolerance for ambiguity, an ability to intuitively understand and accurately interpret situations, and an ability to identify and frame a problem accurately, then it is wisdom as well as intelligence that is being valued. If you are interning in an agency where empathy, concern, insight, and efficient coping skills are valued, again wisdom as well as intelligence are being valued (Hanna & Ottens, 1995). Ideally, your supervisor possesses both qualities—intelligence and wisdom—and has modeled the importance of both for you.

Wisdom is what is beyond textbook. It is what makes a supervisor empowering, it is what makes a leader remarkable. The differences between these two modalities reflect the differences between what can be instilled and developed in you (intelligence) and what you must learn to develop through experience and commitment to a more visionary way of dealing with issues (wisdom).

The third area of awareness that interns tend to develop in the Competence stage has to do with *professional behavior*. This awareness tends to take the form of curiosity about politics and staff behavior, and their effects on the quality of the work. We address this area of awareness in the next chapter, where we review the ethical and legal issues of the workplace that are relevant to interns.

SUMMARY

The rewards of this long journey in experiential learning are finally realized when you enter the Competence stage. Although hardly without its concerns and challenges, this stage is one in which you can indulge yourself and enjoy the feelings of finally reaching your goals. The transformations in development that you have experienced and the sense of empowerment that you have developed are evident not only to you in how you feel and go about your work, but are evident to those around you as well.

As you continue to develop competencies and a sense of professional identity, you begin to realize that you have concerns about situations, behaviors, or politics that, up to this point, either did not seem evident or did not carry the importance they seem to now have. Your status has become that of an emerging professional, a consequence of which is an enhanced awareness of the ethical and legal issues that are part of the workplace. Those are the issues we look at in the next chapter.

For Further Reflection

FOR PERSONAL REFLECTION:
SELECT THOSE QUESTIONS MOST MEANINGFUL TO YOU

1. Take a moment to think about just how fulfilled you are with your internship. What aspects of your experience contribute to this sense of fulfillment? Which aspects tend to interfere with it?

2. On a piece of paper, jot down all the factors you can think of (including the ones mentioned in this chapter) that are important to your internship (positive and negative). Then, next to each one, indicate the amounts of "too much," "too little," or "just right." Chances are that what's just right leaves you with the greatest feelings of satisfaction.

SPRINGBOARDS FOR DISCUSSION

As you read over the following questions, think about the ways in which your perspectives about professionalism have changed during the internship. Share them with your peers and discuss your differences. What do you think contributes to these differences?

1. What is your operating definition of professional? What behaviors and values do you associate with being professional?

2. What professional behaviors are the norm for your field site? What behaviors contradict the norms? What values are reflected in both instances?

3. Think about someone you met in the course of your internship who meets your definition of professional. What is it about this individual that you find particularly admirable?

4. How do you see yourself measuring up to your definition of professional? (Be as specific as you can be.)

For Further Exploration

Baird, B. N. (2002). *The internship, practicum, and field placement handbook.* Upper Saddle River, NJ: Prentice Hall.

Comprehensive and covers a wide range of topics relevant to the clinical/counseling dimension of human service work.

Jackson, R. (1997). Alive in the world: The transformative power of experience. *N.S.E.E. Quarterly,* 22(3), 1 & 24–26.

Explores what it is about experiential education that "gives experience the power to transform" an individual's thinking.

References

Chickering, A. W. (1969). *Education and identity.* San Francisco: Jossey-Bass.

Collins, P. (1993). The interpersonal vicissitudes of mentorship: An exploratory study of the field supervisor-student relationship. *Clinical Supervisor, 11*(1), 121–136.

Dreyfus, S. E., & Dreyfus, H. L. (1980). A *five stage model of the mental activities involved in directed skill acquisition* (Unpublished Report F49620-79-C-0063). Air Force Office of Scientific Research (AFSC), University of California, Berkeley.

Erikson, E. H. (1963). *Childhood and society.* New York: Norton.

Hanna, F. J., & Ottens, A. J. (1995). The role of wisdom in psychotherapy. *Journal of Psychotherapy Integration 5,* 195–219.

Hersey, P., & Blanchard, K. (1982). *Management of organizational behavior: Utilizing human resources* (4th ed.). Upper Saddle River, NJ: Prentice Hall.

Huber, R. M. (1971). *The American idea of success.* New York: McGraw-Hill.

King, M. A. (1988). *Toward an understanding of the phenomenology of fulfillment in success.* Unpublished doctoral dissertation. University of Massachusetts, Amherst.

Kolb, D. A. (1984). *Experiential learning: Experience as the source of learning and development.* Upper Saddle River, NJ: Prentice Hall.

Lamb, D. H., Baker, J. M., Jennings, M. L., & Yarris, E. (1982). Passages of an internship in professional psychology. *Professional Psychology, 13*(5), 661–669.

Mozes-Zirkes, S. (1993, July). Mentoring integral to science, practice. *APA Monitor,* 34.

Tentoni, S. C. (1995). The mentoring of counseling students: A concept in search of a paradigm. *Counselor Education and Supervision, 35*(1), 32–41.

Tryon, G. S. (1996). Supervisee development during the practicum year. *Counselor Education and Supervision, 35*(4), 287–294.

CHAPTER 13

Considering the Issues: Professional, Ethical, and Legal

> *I have to make sure that I am not in the wrong place at the wrong time. I have to be careful . . . because I don't want to be in a bad situation. I try to stay out of situations where I could be forced to make a bad decision.*
> STUDENT REFLECTION

A s you go about working independently now, making decisions and testing the limits of your own competence, you will encounter situations about which you will feel uncomfortable. These are situations that will give you cause to pause and think. You may have already found yourself feeling this way and were unsure about what conclusions to draw, what decisions to make, or what actions to take. Welcome to the world of professional work and its many challenging issues.

Some of these issues have ethical dimensions; some have legal dimensions; and some have a combination of both dimensions. This chapter will help you develop a way of thinking about these issues and identify tools and resources you can use to deal with these matters in the course of your work in the field.

We have one reminder before we move forward with this discussion. Although many readers using the text are interns, not all are. Many students are conducting practica, co-ops or service-learning experiences. The academic level of your field experiences also varies considerably, from two-year undergraduate programs through doctoral programs. Although this text is intended primarily for the helping professions, there is a wide variety of professional roles to consider: human service professionals; social workers; mental health, school guidance, school adjustment, vocational, rehabilitation, and pastoral counselors; and psychologists. In addition, many of you are in public service placements such as probation, parole, corrections, and law enforcement. Consider, too, that your academic majors vary considerably, including English, Communications, Nursing, Business Administration, Social Justice, Sociology, and Social Work. Consequently, as you move along in the chapter, keep in mind that at times, there will be content of significant relevance to what you are doing because of the similarities across all these areas; and at other times, you will need the resourcefulness and guidance of your instructor to access information relevant to your specific field situation.

REALIZING THE ISSUES

In an effort to provide effective services, there has been a growing need since the late 1980s for helping professionals to know about their rights and responsibilities, as well as those of others in the helping process. Although a detailed discussion of these issues is beyond the scope of this chapter and text, a useful way of organizing the most common issues is to think of them in terms of the distinct aspects of the work you do on-site (Chiaferi & Griffin, 1997). We use this framework with some adaptation and incorporate issues across the spectrum of the helping professions (American Psychological Association, 1992; Baird, 2002; Chiaferi & Griffin, 1997; Collins, Thomlison, & Grinnell, 1992; Corey et al., 2003; Goldstein, 1990; Gordon, McBride, & Hage, 2001; Martin, 1991; Schultz, 1992; Wilson, 1981).

The issues are grouped into four aspects of the work: professional issues, integrity issues, intervention issues, and internship issues. The last category, internship, focuses on your primary role in the field experience, and we don't let you lose sight of the legal and ethical aspects of that role as you assimilate into the profession.

In each of these categories, we identify situations that you may face in your internship. Some of the situations overlap categories, due to the nature of the work. We have listed them in detail so you can see the vastness of your responsibilities and the potential issues you could face. You are expected neither to remember them nor to know them, though in time and through practice they will become quite familiar to you. At the very least, you will know where to go to learn more about them. The issues are listed by category in Table 13.1. For the most useful and comprehensive treatment of this subject, we refer you to the seminal text *Issues and Ethics in the Helping Professions* (Corey et al., 2003).

Professional Issues have to do with how you engage your profession, and include such issues as educational preparation, diversity awareness, and dressing for the role. An example of a dilemma that falls into this category is your having information about a staff member who is demonstrating insensitivity to the culture of a client's family, is misrepresenting her qualifications to work with such families, and is now being promoted to supervisor of your work group at the site. It happens that you and the client share a similar cultural identity. It also happens that you and this staff member have had difficulties working together in the past. You are wondering whether you should disclose your concerns to your site or faculty supervisors, talk directly with the staff member, or say nothing.

Integrity Issues have to do with how you approach your work on a daily basis and include such issues as confidentiality, disclosure, and record-keeping standards. An

TABLE 13.1
Issues Facing Interns in the Helping Professions

Professional Issues

- Competence in doing the work
- Frequency and focus of supervision
- Consultation
- Education
- Diversity awareness
- Grievance issues
- Limitations in the scope of practice
- Credentialing/license standards and requirements
- Advertising for services
- Dressing for the role
- Relationships with supervisors and staff
- Managing the risks of physical danger and legal liabilities

Integrity Issues

- Dual/multiple status relationships
- Obtaining information
- Disclosure of information
- Record keeping
- Informed consent
- Privileged information
- Right to privacy
- Confidentiality
- Upholding the values of benevolence, autonomy, nonmaleficence, justice, fidelity, veracity

- Exceptions to confidentiality, including abuse/neglect cases
- Dangerous client cases (self and others)
- Third-party payer requests
- Responses to court orders
- Release of information to clients
- Duty to warn and protect
- Integrity of clients
- Attraction/intimate relationships (emotional, physical, sexual)

Intervention Issues

- Clinical issues (transference and countertransference)
- Limitations on scope of responsibilities
- Client's right to self-determination
- Management of referrals
- Size and nature of caseload
- Termination
- Working with special populations
- Abandonment by therapist
- Obtaining and releasing information
- Sharing of information with colleagues
- Emergency response during nonworking hours
- Differences in legal and ethical practices
- Individual vs. group vs. marriage and family interventions

example of a dilemma in this category is your being out to dinner and overhearing the conversation of an agency worker who does not know you, but whom you recognize. The conversation is about a client, and the worker identifies enough data that you recognize the case from a staff meeting earlier that week. You know from your orientation period that the agency's policy states that workers are not supposed to talk about cases outside of the office, or in the very least, they must not disclose any identifying information about the case. The person with whom the worker is speaking is your

TABLE 13.1

Issues Facing Interns in the Helping Professions *(continued)*

Internship Issues

- Right to quality supervision

- Responsibility to confront situations in which educational instruction is of poor scholarship and nonobjective

- Disclosure of risk factors to all potentially affected parties (to site supervisor about intern; to intern about site supervisor)

- Behavior consistent with community standards and expectations

- Awareness of risk status of agency

- Active involvement in the placement process and consideration of more than one placement site

- A clearly articulated learning contract that identifies mutual rights, responsibilities, and expectations

- A service contract with the agency that defines the limitations of the intern's role

- Liability insurance

- The prior knowledge clause

- Assurance of work and field site safety

- Assumption of risk as limited to ordinary risk

- Employer-employee-independent contractor relationship

- Compensation: stipend, scholarship, taxable/taxfree

- Deportment

- Negligence

- Malpractice

- Implication of federal funds and related statutory and regulatory requirements

- Use of college work-study funds for interns

- Grievance processes

- Informed consent in accepting an internship

- Respect for the prerogatives and obligations of the institution

- Responsibility to confront unethical/illegal behaviors

- Public representation of self and work

- Disclosure of status as intern

- Boundary awareness

- Boundary management

- Personal disclosures

- Criminal activities

- Political influences/corruption

- Subversion of service system

- Office politics

child's teacher. Neither of them saw you, and you are not sure what to do. You realize that something inappropriate has occurred, but you are also hoping to be hired at the end of your internship and are hesitant to pursue the issue for fear of making the wrong move.

Intervention Issues have to do with working directly with clients and include such matters as managing clinical issues, making referrals, and overseeing your caseload. An example of a dilemma that would fall into this category is finding out at an Alcoholics Anonymous meeting, which you attend for personal reasons, that your supervisor's client is planning to leave the country in the next couple of weeks. Of particular concern is the fact that a friend told you that she overheard the client threatening to hurt a former girlfriend. You had been told that what goes on at Alcoholics Anonymous meetings is confidential, and you know how strongly confidentiality is valued in your future profession. You wonder how to uphold your responsibilities to all the parties involved (the client, the girlfriend, AA, and your profession), many of which seem to be in conflict, and how to determine which responsibilities take priority over others.

Internship Issues have to do specifically with your role as an intern and issues of academic integrity, competence, and supervision. *Academic integrity* issues include a quality field site, responsible contracts, and a seminar class that ensures a "safe place" for reflective discussions (Rothman, 2000). *Competence* issues include knowing your limitations and finding a balance between challenging work and a realization you have exceeded your level of competency. It is important that you know the limits of your skills and seek help as needed. (Gordon, McBride, & Hage, 2001; Taylor, 1999). *Supervision* issues include the assignment of an appropriate supervisor who knows how to supervise interns in particular and can appropriately deal with such complex issues as client abandonment, the dynamic of attraction in the supervisory relationship, and quality evaluations of the intern.

The potential for liability here is very real. Interns may deal with all the issues identified in the previous three categories and are held to the same standards and ethical documents as employees. In addition, there are a number of issues that are specific to experiential education and internships in particular. For example, as an intern, you have the multiple roles of being a student and a "consumer" of services by virtue of your need for supervision. If you are in the helping professions, then you have the additional responsibility (role) of being a care provider to others. In that case, accountability is demanded in all three areas.

An example of a dilemma in the internship category is your becoming aware of unethical and possibly illegal practices at a site where you have been offered a paid field experience. The site is out of state and is not one of your campus's regularly used agencies. Your family is relocating to that state for financial reasons, and having a paid internship would help prevent further hardship on your family. You doubt that the field placement coordinator for your academic program is aware of the improprieties at the agency. And you are questioning whether to inform the campus about what you know or be silent and help your family.

ETHICAL ISSUES: A WORLD OF PRINCIPLES AND DECISIONS

You might be wondering how you have managed to survive for so long without knowing about these issues! This is exactly how our students feel after studying them in a semester-long course.

Chances are that your basic values have held you in good stead. However, the list can be overwhelming at first, even after several readings. Most likely, there are many issues about which you never heard and many that seem remotely familiar. Reading through them is a good start. Now that you are becoming familiar with the language of ethics, even if you do not yet know all that it entails, it is a good time to go ahead and think about the dilemmas that already have surfaced in your internship and how satisfied you are with your responses to them.

Talking the Talk

In order to have a useful discussion about these issues, there needs to be a shared language for communicating and a common understanding of the problem. In the discussion that follows, we identify some of the most central and frequently used terms related to ethical matters and their working definitions. It might take you a number of readings to become familiar with their meanings, so do not be concerned if at times there seems to be too much to grasp. You will become comfortable with the material, but it will take time.

We'll start with the term issue. An *issue* refers to a point that is in question or in dispute, and *professional* refers to matters that pertain to an occupation. *Professional issues*, hence, refer to some aspects of how one goes about doing one's work that have become a matter of debate among others. *Standards*, when used generically, refer to guidelines or codes that govern the behavior of members of a given profession. *Ethical* describes someone acting in accordance with professional standards, codes, guidelines, or policies, and *unethical* suggests that they are not. *Legal* describes someone acting in accordance with the law, and *illegal* suggests that they are not.

Besides these terms, which are related directly to the responsibilities of a profession, the terms *values, moral,* and *ethics* are very important to any discussion about ethical, legal, or professional issues. *Values* refer to what is intrinsically good, useful, and desirable; *moral* refers to what is right or wrong conduct in its own right, based on broad mores such as religious principles; and *ethics* refers to the moral principles or rules of conduct of a particular group, such as human service ethics, criminal justice ethics, mental health ethics, and so on (Corey, Corey, & Callanan, 2003; Pollock, 1998).

Rules of the Trade

In addition to a common language, it is important to have access to common resources for your profession. For one thing, although your understanding of the issues in the helping professions has just begun, you are still responsible for acting in accordance with the values and standards of the profession. These values and standards are embodied in ethical documents variously referred to as *guidelines, standards, regulations, policies,* and *codes.*

It is precisely these types of documents that are reviewed when conflicting or questionable situations arise. Unfortunately, they don't necessarily lend themselves to clear responses, interpretations, or resolutions of ethical or professional conflicts. However, they do provide guidelines for behavior and discussions. Technically speaking, there are differences between these documents. If you are curious about these differences, read on. Otherwise, move ahead to the next part of the chapter, where we discuss what can help you to make effective decisions when these documents fail to adequately guide you (see "Beyond the Documents" on page 233).

You wanted to know, so here goes. It would be helpful to your understanding of the discussion that follows if you had a copy of the ethical document that guides the professionals in your agency. If you can find one, begin by taking the time to notice how the document is titled. If it reads *guidelines*, the statements reflect recommendations from professional groups that describe acceptable behaviors for the profession and its practitioners. If the document reads *standards*, the statements reflect the rules of behavior for the profession, drawn up by members of the profession itself, often carrying civil sanctions and setting the parameters of ideal behaviors. If you have heard your supervisor or field instructor talking about accreditation teams or visits, they are referring to an assessment of the site or academic program based on designated standards of behavior for the profession.

If, on the other hand, the document reads *regulations*, the statements refer to dictates typically from governmental authorities and often specify sanctions for not complying with them. A *policy* refers to the procedures or courses of action set forth by an organization to ensure expediency and prudence in the work getting done. Most organizations have policy manuals for their employees and interns to read as part of their orientation to the work and workplace.

Finally, if the document you are perusing reads *code*, its statements reflect the beliefs about what's right and correct conduct for a profession. Codes often include standards of practice along with statements that embody the values of a profession. Like the learning contract, codes tend to be living documents that continually evolve. They promote professional accountability and facilitate improved practice by protecting the professional from ignorance (for example, from malpractice suits so long as the professional acts in accordance with acceptable standards); protecting the public from the profession (that is, protects the consumer from harm); and protecting the profession from the government (that is, the profession governs and regulates itself, protects itself from internal struggles, and establishes agreed-upon standards of care) (VanHoose & Kottler, cited in Bradley, Kottler, & Lehrman-Waterman, 2001).

If you have not already done so, ask your supervisor for a copy of the ethical document that guides your coworkers, as that is the document that also guides you. We suggest that you take the time to peruse it. Some of you will be looking at a two-page document and others will be faced with 20 pages of professional rules! Reading the shorter documents is a reasonable exercise. However, if you are bound by the American Counselor Association's (ACA) code or that of the American Psychological Association (APA), reading such lengthy documents right now is not a reasonable undertaking. What is expected of you, though, for the purposes of the field experience, is to become familiar with the categories of codes and those categories relevant to your specific field experience. Bring the document to seminar class. Unless you and all your peers are in the same specialty in the helping professions, for example, mental health work, the documents you bring to class will be varied in specialty area, length, and intent.

We include two documents as appendices to this chapter that focus on standards of practice: one for experiential learning *(Eight Principles of Good Practice for All Experiential Learning Activities)* and the other a generic one for the helping professions *(Ethical Standards of Human Service Professionals)*. These documents are well worth reading, as they can inform your expectations of the field experience (see the box titled "Standards of Practice and Ethics").

It is important that you be familiar with the resources of your future profession. Your site supervisor can be very helpful in identifying them for you. For example, a very helpful resource is the book *Codes of Ethics for the Helping Professions*, which includes

---

Standards of Practice and Ethics

There are two sets of standards included in the Appendix to this chapter that will help clarify responsible ways to carry out the work of the profession and your internship. One is the *Standards of Practice: Eight Principles of Good Practice for All Experiential Learning Activities* developed by the National Society for Experiential Education (NSEE). The other is the *Ethical Standards of Human Service Professionals* developed by the National Organization for Human Service Education (NOHSE) and the Council for Standards in Human Service Education (CSHSE). This document is a generic set of broad guidelines for the many disciplines across the human service profession. We hope that you take the time to actually read through these documents before continuing your reading. The standards will clarify responsible ways to carry out the work of the profession and your internship. They also will lend considerable insight to the next part of the chapter, where we identify many of the issues that could potentially become problematic for you or any practitioner.

---

the full text of documents for 15 professional specialties (Wadsworth Group, 2003). In addition to the documents, you will want to have the addresses (electronic and postal) for the professional organizations. Membership in such organizations can be expensive, but well worth the fee; most have a student category of membership with adjusted fees. The vast majority of national organizations are replete with information and resources for their members. In addition to a variety of support services, such as casebooks and libraries of instructional videos, many offer legal counseling and services as necessary.

Beyond the Documents

Regardless of how detailed the documents may be or how many times you read them, there may be no answers forthcoming to help you resolve the conflict you experience. When ethical rules fail to provide a direction toward a solution, then the *ethical principles* of the profession can be taken into consideration to guide your decisions. Ethical principles are fundamental doctrines of the profession rooted in common sense morality.

There are six ethical principles that reflect the highest level of professional functioning in the helping professions (Corey, Corey, & Callanan, 2001, p. 15). The principles are based on the works of Kitchener (1984) and Meara, Schmidt, and Day (1996) and probably look very familiar to you. You may not have realized, though, that these principles have more than one purpose: they can guide your work as well as your decision making. These six principles are autonomy, beneficence, justice, nonmaleficence, fidelity, and veracity. If you are not in the helping professions, you may want to locate those principles that guide work in your field and spend some time thinking about them.

Because you probably know these principles well, we will keep our discussion brief and begin with the principle of *autonomy*, which refers to the freedom clients have to control the direction of their lives by making decisions that reflect their wishes; this principle affirms the self-determination of the clients you serve. The *beneficence* principle refers to your commitment as a helping professional to promote "good," as demonstrated by carrying out in your work with competence and without prejudice. This principle affirms the dignity and promotion of the client's welfare. *Justice* refers to your treating others with fairness, regardless of their gender, ethnicity, race, age, religious affiliation, sexual orientation, disability, religion, cultural background, or socioeconomic status. This principle affirms equity in the work that you do. The principle of *nonmaleficence* refers to your commitment to avoid doing harm to clients; this principle affirms respect for them. *Fidelity* refers to your creating a trustworthy relationship with clients by making honest promises and honoring your commitments to them; this principle affirms the right of clients to informed consent before committing to interventions. *Veracity* refers to your being truthful in your dealings with clients; this principle affirms clients' rights to full disclosure. When using these principles to guide your decisions, it is important to be as honest with yourself as possible. Such authenticity will prove invaluable to the quality of your decision making.

RECOGNIZING THE ISSUES IN THE PROFESSIONAL WORLD

Once your concerns focus on developing competence, it is quite common to pay attention to the professional conduct of your coworkers as well as your own. You may not even be aware of it when you begin to notice others' behaviors. For example, you may find yourself paying a lot of attention to the ways staff members go about their work, deal with clients, or conduct themselves with colleagues. Perhaps you are beginning to look at others not just as professionals with roles, but as professionals with moral, ethical, legal, and professional responsibilities to the profession, to clients, to the organization, and to you. You may even be tuning into the subtleties of their behaviors and becoming aware of possible improprieties in how they go about their work.

Questioning Others' Conduct

The improprieties that you observe did not begin when you first noticed them. In all likelihood, it is you who have changed: You can now see what has been there all along. The behaviors and attitudes were just not within your sphere of awareness before. There are a number of possible reasons for this change. First, staff tend not to disclose questionable or surreptitious aspects of themselves so readily to interns or new employees, but rather tend to act as they are expected to in their roles (Kanter, 1977). Consequently, they tend to shield questionable behaviors and attitudes from interns until they get to know them better. Another explanation is that you have been so busy with your increasing responsibilities that you have not had time to notice these behaviors or even think that there was something to notice. A third possibility is that you held stereotypes that needed to change before you were willing or able to see situations for what they were.

There are a number of situations that lend themselves, conditionally or not, to questionable behaviors. Some situations are so obvious that mentioning them seems absurd. However, they do need to be mentioned, because aspiring professionals, like seasoned professionals, are people too, and they have personal frailties that at times compromise their ethical standards. We both have known and worked with individuals who committed improprieties that neither we nor they ever would have expected. The following list (in part from Royse, Dhooper, & Rompf, 2003) mentions some of the more common of these improprieties:

- Being sexually intimate with clients or supervisors
- Taking libelous or slanderous actions against clients
- Behaving in a threatening or assaultive manner against clients or coworkers
- Misrepresenting one's status or qualifications
- Abandoning a client in need of services
- Failing to warn and protect appropriate parties in the case of a violent client
- Failing to use reasonable precautions with clients dangerous to themselves
- Being dishonest or fraudulent in your actions

A word of caution is needed here. Regardless of the reason, you are bumping up against issues now that did not concern you in the past. And, they can become complicated enough to be potential pitfalls for any aspiring or even seasoned professional. When the issues are of an ethical or legal nature (we discuss the legal aspects in the next part of the chapter), the situation needs to be managed with reason and sensitivity. What is at stake is at least someone's feelings and at most someone's career, life, or family. For now, it is important to realize that when the stakes are that high, they are high for you, too. Your opportunities for employment and the future of your career could be compromised if your concerns are not justified or handled professionally.

Questioning Your Own Conduct

When it comes to your own behavior, the important issue is whether you would recognize an ethical issue when dealing with it in the field. For example, if a client of yours wanted to deliver daily newspapers to your office (what a convenience!) or relay some gossip about another client (how tempting), would you see a problem with either of these situations? We find that the areas of *relationships* and *confidences* are particularly challenging for our interns. Perhaps they are for you also.

RESPECTING RELATIONSHIPS

When an intern is engaged in more than one relationship (role) at the same time (or sequentially) with a client, the situation that develops is referred to as *dual* or *multiple relationships* (Corey, Corey, & Callanan, 2003). The nature of these roles can vary from being an acquaintance, friend, or business client to being an intimate, whether emotionally, physically, or sexually (Dorland, cited in Malley & Reilly, 2001; Gordon, McBride, & Hage, 2001; Royse, Dhooper, & Rompf, 2003). The partners in these relationships vary as well, from professors to staff, supervisors, clients, and coworkers

(Malley & Reilly, 2001). These relationships are fraught with complex legal and ethical issues. If you are in one of these relationships, it is best to seek consultation immediately. Another relationship that can be difficult to manage is the *collegial* one, which refers to your relationships with staff, peers, and supervisors involved in your internship. These relationships need to be managed so that professional obligations are upheld. You may want to take a moment to think about what you could do and what you would do if your commitment to the profession was being compromised by how you were handling one of these relationships (Rothman, 2000). Again, if you are in a compromising relationship, it is best to seek consultation immediately.

SAFEGUARDING CONFIDENCES

The other area that tends to challenge interns is that of *confidences*. This issue is rather pervasive and involves *clients' rights* (to privacy, privileged communication, and confidentiality) and *information disclosure*. These matters are likely to arise in your role as an intern, so it is important that you understand them. Depending on the laws of your geographic area, these matters could be legal, ethical, or both.

The clients rights that you deal with on a daily basis probably include privacy, privileged communication, and confidentiality. *Privacy* refers to the constitutional rights of your clients to decide when, where, and how information about them is disclosed to others by you (Corey, Corey, & Callanan, 2003; Eisenstat, personal communication, 2003). The matter of *privileged communication* is a legal concept, and the right to such communication belongs exclusively to the client. This concept protects your client from the forced disclosure of information in legal proceedings (Corey, Corey, & Callanan, 2003). You need to check with your supervisor to determine whether or not clients who see you in a helping capacity retain their rights to privileged communications or lose them because of your status as an intern. If the client does not have the right when working with you, the client needs to know that, and your supervisor needs to guide you in making that happen. *Confidentiality* is a legal, professional, and ethical matter that protects the client in a therapeutic relationship from having information disclosed by you without explicit authorization (Corey, Corey, & Callanan, 2003). Your professional status of intern also has direct implications for whether the clients you work with are protected by confidentiality statutes. Again, you need to discuss this matter with your supervisor, and your clients need to know accordingly.

Disclosure of information can become complicated when you are disclosing information to your colleagues (while on-site or socially) or to your peers (in the seminar class). In both instances, you have a responsibility to the privacy rights of the client, and in the very least, to be familiar with and follow the policies of the agency, disguise all identifying data, and give considerable thought to the question "Why do I have a need to tell this information to someone?"

The issue of disclosure in the world of the helping professional also applies to cyberspace and the information highway. Be it e-mail, chat rooms, bulletin boards, or supervision online, interns need to have a working command of this aspect of their workplace. The electronic obtaining and releasing of information, along with maintaining such information for the agency, can be fraught with potholes for the intern. Whether you are working with databases or releasing information, it is important that you make informed decisions and do so carefully. The agency has its policies to safeguard the maintenance of information, as well as forms to ensure that all criteria are met before information is released by you. If you have not yet reviewed these policies and forms, this is a good time to do so. Also, remember that when you use e-mail, a chat room, or a bulletin board to discuss clients or other sensitive issues, those means of communication are not always secure.

LEGAL ISSUES: A WORLD OF LAWS AND INTERPRETATIONS

An important part of making an ethical decision is knowing the laws that are relevant to and affect your work.[1] Some of you are developing a familiarity with the law, especially if you are interning in legal settings or your work is closely directed by statutes and legal guidelines. Others of you may know little about this aspect of your work. In this part of the chapter, we offer you a way of thinking about legal matters that will help you make better sense of this aspect of your field experience.

For those of you interning in the criminal justice system, legal mandates govern much if not all of your work. For those of you working with dependent individuals, such as minors, elders, and those with special needs, the intent and extent of your work is also largely affected by legal statutes, especially in protective work (abuse, neglect, and exploitation). If you are interning with a legislature, advocating for clients in class action suits, interning in a hospital, human resource department, or in mediation services, you are working with laws. For that matter, all interns are affected to some degree by laws in their work.

A number of legal issues are particularly relevant to interns who work directly with clients. Most of these issues also have ethical dimensions (see Kiser, 2000). Such issues include but are not limited to liability and malpractice; confidentiality, privileged communication, and privacy; disclosure of information; end-of-life decisions; consultations with specialists; crisis intervention; suicide prevention; termination of interventions; intimacy with clients; duty to protect intended victims from violence; informed consent (Berg-Weger & Birkenmaier, 2000; Kiser, 2000). The list does not stop here, but we will.

Your responsibility is to know the legal basis for your agency if there is one, the laws that affect and govern your work, and the ways in which you are bound by those laws in carrying out your responsibilities (Berg-Weger & Birkenmaier, 2000; Gordon, McBride, & Hage, 2001). The best way to learn about these matters is to bring your questions and concerns to supervision. The following paragraphs will help prepare you for those discussions.

Talking the Talk

As was the case with ethical matters in field work, there is a terminology specific to legal matters that you need to know. Again, we will take some liberty and use working

---

[1]Appreciation is extended to Professor Steven Eisenstat of Suffolk University Law School, who specializes in civil tort law, for his assistance with the section of this chapter on legal issues.

definitions where possible so that you have a sense of the language and its implications. We know that the information here is typically what interns most want to know about legal matters. However, the information is very technical, and it is not possible to define it without a great deal of detail. So, on the one hand we risk oversimplifying a complex body of information, and on the other hand we risk boring you or causing you undue concern. We will do our best to choose a middle ground. We advise you, though, throughout this discussion and throughout your internship, to bring all matters to supervision if you do not have a working understanding of them.

We'll start with the term *tort*. A tort is a civil wrong or injury done to another that is not based on an obligation under a contract. There are three types of torts: intentional torts, negligence torts, and strict liability torts. For the purposes of your internship, it is the *negligence tort* that is of most concern to you, your site supervisor, and your campus instructor. For an act to be a negligence tort, there must be a legal duty, owed by one person to another, a breaking (*breach*) of that duty, and harm caused as a direct result of the action. For example, if you are interning at a home health care agency and you voluntarily assume responsibility for an elder in that community, you then are in what is referred to as a *special relationship* with that person. Your duty to your client would be considered breached if you fail to provide the standards of care of the home health care profession. You could do this either by failing to take certain required actions or, if you did act, doing so in a way that did not reflect the standards of care for the home health care profession. It makes sense to raise the issue of *breach of duty* with your supervisor so you can better understand how you could be at risk for such lawsuits (Eisenstat, personal correspondence, 2003).

Negligence torts, then, can result when you fail to exercise a reasonable amount of *care* (standard of care) in a situation that causes harm to another person or to a thing. The basis for the negligence tort can involve doing something carelessly or failing to do something you are supposed to do.

There are two forms of negligence: ordinary negligence (that is, failing to act as a reasonable person would) and aggravated negligence (that is, reckless or willful behavior). It is the *ordinary negligence* tort that is the more likely concern in your internship. An example of ordinary negligence occurs when police, probation, corrections, and parole officers fail to perform duties owed to their charges, or when they perform duties inadequately (Eisenstat, personal correspondence, 2003). For example, correctional officers have a duty to check regularly on the inmates under their care. If a correctional officer fails to do so and an inmate commits suicide, the officer could be found negligent in terms of his or her supervision responsibilities. So, it makes sense to raise the issue of negligence with your supervisor and the potential pitfalls you face so that you can better understand how you could be at risk for such lawsuits.

The term *malpractice* is one you are sure to have heard and know enough about that you do not want it to be a part of your field experience! Malpractice, which is a form of ordinary negligence, refers to lawsuits brought for an act that you perform in your professional capacity. This type of lawsuit involves professional misconduct or unreasonable lack of skill on your part that results in injury or loss to your client. For example, if you are an intern at a residential facility for emotionally disturbed adolescents, and you fail to take such adequate precautions that are ordinarily provided by other residential facilities or workers in the profession, and your actions or lack thereof result in one of the residents committing suicide, then you most likely would face a malpractice lawsuit (Eisenstat, personal correspondence, 2003).

Although not reassuring, it is important to know that there are situations that tend to increase your liability for a malpractice lawsuit. (Being *liable* or having a *liability* means a breach of duty or obligation to another person.) For example, if you fail to use acceptable procedures, or you break a contract, or you use interventions for which you were not trained, or you fail to choose the most helpful interventions, your risk of liability increases significantly. And, it does not stop there. If you fail to warn others about or protect others from potential danger, or you fail to secure informed consent appropriately, or you fail to disclose to your client the possible consequences of services and interventions, then your risk of liability could increase. Again, it is important to bring the issue of malpractice to supervision so you can better understand how you could be at risk for such lawsuits.

Rules of the Trade

In addition to having an understanding of the terminology, it is important to have an understanding of the legal framework. Let's start at the beginning with laws. A useful way of thinking about laws is how they are classified. For example, laws can be classified according to how they come into existence. Laws in the United States derive from our constitutions (*constitutional law*, from state and federal constitutions), our legislatures and governmental agencies (*statutes* and *regulations*, respectively), and our *common law* (*case law* from prior decisions by trial courts or appeals courts). Another way laws are classified is by the nature of their intent: criminal or civil. *Criminal law* refers to a group of laws that seeks to resolve disputes between the government and people. Criminal law seeks punitive measures such as imprisonment and fines to right a wrongdoing (Eisenstat, personal correspondence, 2003). Many interns work with criminal law on a daily basis, namely, those in criminal justice settings, legal offices, domestic violence agencies, and protective work with dependents.

Civil law, of which tort law is an example, on the other hand seeks to resolve disputes between people by enforcing a right or awarding payment or what is referred to as *damages*. Its primary intent is to repair rather than punish behavior, as is the intent of sanctions for criminal matters (Eisenstat, personal correspondence, 2003). An aspect of civil law that many interns work with is mental health law. This body of law regulates how the government takes care of or responds to people with mental health challenges. If you are interning in a mental health clinic, hospital, residential setting, or community shelter, your work is affected by this body of law. In some instances, students deal with both civil and criminal law. For example, interns at offices of the American Civil Liberties Union deal with constitutional rights, which deal extensively with civil law as well as criminal law issues.

Another way of thinking about law and the helping professions is to separate the laws according to the aspects of the work (Horejsi & Garthwait, 2002). For example, there are laws that regulate the *services or actions* that you can give to a client. There are laws that regulate the *work of the human service agency*, such as working with youth, working with elders, and working with mentally ill individuals. And there are laws that regulate the *professional practice* of the professional, such as deportment issues, licensing issues, and issues such as confidentiality and informed consent.

Relevant Legal Matters

The two legal matters of most relevance for students in field placements are grouped in the general categories of *standards of care* and *supervisory liabilities*. Again, these are complex areas of inquiry, and it is not possible for us to address them adequately in one chapter on ethics and laws. However, we hope to give you a way of thinking about them so that you can bring the matters to supervision and become better informed in these important areas of practice. We base the discussion that follows on the work of a number of writers and experts in this field, including Berg-Weger and Birkenmaier (2000), Champion (1997), Corey, Corey, and Callanan (2003), Eisenstat (2003, personal correspondence), Falvey (2002), Gordon, McBride, and Hage (2001), Horejsi and Garthwait (2002), Malley and Reilly (2001), Oran (1985), and Pollock (1998).

STANDARDS OF CARE

Interestingly enough, an area of legal matters not universally and directly addressed in the ethical standards for helping professions is that of *reasonable standard of care* (Kiser, 2000). This is a matter that is both ethical and legal in nature and one that affects your work in the field every minute of the day. Kiser (p. 122) has defined the components of a reasonable standard of care to include, but not be limited to

- Knowledge of the clients and services being given
- Delivery of services and interventions based on sound theoretical principles
- Reliability and availability of services to clients
- Initiative and acting on behalf of client and public safety
- Adherence to ethical standards of the profession in relation to client care
- Systematic, accurate, thorough, and timely documentation of client care

SUPERVISORY LIABILITIES

The second legal matter has to do with *supervisory malpractice*. In this instance, it is the behavior of the supervisor that comes under legal as well as ethical scrutiny. As you may be aware, your supervisor has liability for your work, because when your supervisor agreed to supervise you, he or she accepted responsibility for all of your work, including your work with clients, and for your behavior (deportment) during the internship. If this responsibility sounds pretty serious, it is.

Failure to supervise the professional staff appropriately has been the cause of a growing number of malpractice suits (Sherry, cited in Falvey, 2002). This type of lawsuit concerns the quality of supervision you are receiving in the field. The legal scrutiny a supervisor faces in such a lawsuit results from alleged negligence in carrying out the supervisory responsibilities and subsequent injury or damages occurring. You, the intern, along with whoever may have been injured as a result of improper supervision, become the *plaintiffs* (that is, the ones who bring the complaint), and your supervisor becomes the defendant in such a lawsuit for negligence. For example, you have been directed to conduct an in-home assessment to determine the removal of a child based on alleged neglect by the parents. In the process of conducting the interview, the mother becomes despondent, leaves the interview, goes into the bathroom, and slashes her wrists. If your supervisor did not prepare you adequately to respond to and manage the range of possible reactions to such an interview, your supervisor's risk for liability for failing to train you adequately increases substantially. Such preparation by your supervisor could include but not be limited to having you observe and/or conduct such an interview under the direct supervision of an experienced worker, or talking with you about the potential for self-destructive reactions to such interviews (Eisenstat, personal correspondence, 2003). In this scenario, the family of the mother could also bring suit against the supervisor. Hopefully you will never have such experiences as part of your internship.

There is also the potential for another type of negligence liability on your supervisor's part, that of *vicarious liability*. Under vicarious liability, your supervisor could be held responsible for your negligence even if your supervisor did not act negligently. For vicarious liability to apply, typically there must be some form of an employer-employee relationship, and you, the employee, must have acted within the scope of the employment. In addition, your supervisor must have the power to control and direct your work. Most likely, these conditions are inherent in the supervisory relationship between you and your supervisor, whether the internship is a paid experience (for example, a stipend) or the internship is part or all of your employment. Using the example of the assessment interview for neglect, vicarious liability would apply even if the supervisor did discharge her responsibilities in a responsible manner and prepare you adequately for the interview. If you are negligent, the supervisor can still be held legally responsible (vicarious liability) for your work (Eisenstat, personal correspondence, 2003).

At this point in your understanding of liability, you may be wondering, under what circumstances does your supervisor compromise his or her liability to you? Four major sources of supervisor liability have been identified by Harrar et al. (cited in Falvey, 2002). We think it's important to identify them because we believe that you have a right to know how your supervisor is responsible to your supervision in the field. First, if the supervisor is *remiss in carrying out supervisory duties* for planning your internship, the direction of your internship, or the outcome of your work, then your supervisor's risk of liability can increase. And that's not all. If your supervisor *gives you inappropriate advice* about a treatment intervention that you use and the intervention is to the detriment of the client, the supervisor's risk of liability can increase. Third, if your supervisor *fails to listen* attentively to your comments about a client and in turn fails to understand the needs of the client, the supervisor's risk of liability can increase. Last, if your supervisor *assigns you tasks beyond your competence*, then the supervisor's risk of liability can increase.

All of these conditions make good common sense, and our experience is that students know intuitively when they are being short-changed or otherwise not being given quality supervision. However, seeing them in print can be most affirming for the intern, as this intern's comments reflect: "I just wished I had known exactly what the supervisor is legally suppose [sic] to do. . . . Then I would have been able to point to something in writing. It was very hard knowing the supervisor was wrong but not having anything to point to and say 'This is what you are suppose [sic] to be doing for me.'" It would have been helpful for this intern to be aware of what Munson (cited in Royse, Dhooper, & Rompf, 2003) refers to as "the rights of practicum students." In particular, the rights that would have been useful for this student to know about include

- The right to a field instructor who knows how to supervise, that is, has been adequately trained and skilled in the art of supervision
- The right to a supervisor who supervises consistently at regularly designated times
- The right to clear criteria when being evaluated, and
- The right to growth-oriented, technical, and theoretical learning that is consistent in its expectations.

GRAPPLING WITH DILEMMAS

One of the most wrenching aspects of working in the helping professions is dealing with a dilemma that involves the welfare of another individual, family, or community. A *dilemma*—be it ethical, legal, or both—refers to a struggle that occurs among alternative courses of action that might resolve a situation. To complicate things, the choices of courses of action tend to be correct in their own right, but they conflict with each other. A dilemma, then, is a situation in which you can find yourself facing more than one justified course of action, that is, two "right" ways of responding.

Our interns tell us that once they develop an understanding of the issues and become comfortable with the language, they begin to see the issues in their daily work. It just so happens that you tend to develop this awareness at the same time that you are moving into a collegial-like relationship with your supervisors. Consequently, you are much more apt to talk about incidents, behaviors, and concerns at this point in your field experience than you were a couple of months ago.

The next hurdle is to recognize a dilemma when you see one, which is no easy feat. Our experience tells us that recognizing dilemmas as such is quite challenging for both undergraduate and graduate students, as well as for experienced professionals. Often, students cannot readily name ethical issues when they see them, and they do not necessarily see them in a given situation. Academic programs in human services education sometimes require a course in professional issues, but more often will offer it as an elective course or cover the content in other course work (M. DiGiovanni, personal communication, May 2003). Criminal justice programs, on the other hand, tend to require such a course, as do graduate programs in counseling psychology. If you have not studied or discussed these issues academically, and you are feeling unprepared for these challenges, it makes sense. Hopefully, this discussion will help you to frame your understanding of the issues and learn what questions to ask.

Identifying Dilemmas

There are three basic types of dilemmas: (a) those that result from your own decisions, behaviors, or attitudes; (b) those that result from another person's decisions, behaviors, or attitudes and directly affect you; (c) those that you observe from a distance, such as what you have seen or heard, but that do not directly affect you. An example of the first type of dilemma is considering whether to engage in a dual relationship with a client, such as knowingly working with a client whose sister you have dated. An example of the second type of dilemma is working with a client who, unbeknownst to you, is your cousin's intimate partner. An example of the third type of dilemma is observing or hearing about a coworker dating a relative of his or her client. These three types of dilemmas can be seen in all four categories of issues described earlier in the chapter in Table 13.1.

Walking the Walk

You certainly will be exposed to ethical issues during your field work. You may even experience an ethical dilemma. If that happens, remind yourself that you are no stranger to facing difficulties and that you have what it takes to work your way through yet another challenge. And, like the difficulties you faced in the past, an ethical issue can become a problem if you do not manage it effectively.

One way of dealing with an ethical issue is to anticipate it. A useful and effective way of anticipating it is to rehearse in your mind the best possible response (how you would like to respond to it), the worst possible response (your worst nightmare), and a more realistic response (found somewhere between these two extremes). However, there are many situations for which this framework will not work, and for such situations, you will need a way of thinking about making ethical decisions.

Thinking critically about ethical situations is important to making responsible decisions about them. Although most of you have heard the term, "critical thinking," throughout your education, it is quite possible that you do not know what skills are necessary to think critically about an issue, a situation, a decision, or an action. To think critically means to use standards of reasoning such as clarity, precision, accuracy, relevance, depth, and breadth. It involves evaluation techniques, weighing alternative perspectives, and genuine efforts to evaluate all views objectively. Examples include such skills as articulating ideas, asking significant questions, problem solving, and openness to contradictory ideas (Alverno College Productions, 1985). Why so much thinking? Thinking thoughtfully and in demanding ways will help you make wise choices, choices that most likely will help your clients reach their goals (Gibbs & Gambrill, 1996, p. 6).

If your field work is that of an emergency responder, such as a child protective services worker, you do not always have time to think. You need to act and act fast. Time to think becomes a luxury that is only afforded during down time at the agency, and that time tends to be allocated to processing paperwork and feelings related to recent crises. However, in the absence of using critical thinking skills, a number of mistakes have been identified that have serious implications for responsible work, such as misrepresenting clients, increasing client dependency, withdrawing or extending interventions inappropriately, focusing on irrelevant factors, or overlooking client's strengths (Gibbs & Gambrill, 1996).

TEN STEPS TO DECIDING DILEMMAS

I find that when I think about the whole situation (ethical decision), I realize that unethical decisions have a lot of repercussions. . . . People lose their jobs and I don't want to get into trouble.
STUDENT REFLECTION

There are many decision-making models to guide you in developing critical thinking skills. Some are specifically intended to deal with ethical or legal matters in the helping professions (see, e.g., Corey & Corey, 2003; Corey, Corey, & Callanan, 2003; Gibbs & Gambrill, 1996; Kenyon, 1999; Neukrug, 2000; Rothman, 2000; Schram & Mandell, 2002; Tarvydas, cited in Tarvydas, Cottone, & Claus, 2003; Woodside & Mc-Clam, 2002).

The model we offer is one that we use when teaching about ethical and legal issues. The model is based in part on the Steps to Creating Change in Chapter 11 and incorporates adaptations of other models as well (Close & Meier, 1995; Corey et al., 2003). The model offers opportunities to think critically in practical ways and to develop a reasoned response to and action plan for the presenting problem (Table 13.2).

1. Name the Problem Collect as much information as you can about the situation. Clarify the conflict. Is it moral? Professional? Ethical? Legal? Given that there are no right or wrong answers to the situation, anticipate ambiguity and challenge yourself to consider the problem from multiple perspectives.

2. Narrow the Focus Once you have gathered as much information as is reasonable, list the issues you are confronting. Some are more important than others. Describe the critical issues and players; discard the unimportant ones.

3. Consult the Documents Review the guidelines/codes/standards of your profession and the policies and regulations of your agency to determine whether possible solutions are suggested. Identify aspects of the codes that apply. How compatible are your personal values with those of the profession? What rationales support those areas of conflict between your personal values and ethics and those of the profession?

4. Consider the Laws Chances are you are just becoming familiar with the laws—both civil and criminal—that are relevant to your work. Even so, consult those laws. Once familiar with them, you can contact your site's legal counsel for advice (your supervisor should be made aware of this first!). And, if your curiosity is not satisfied with the information you get from legal counsel, you can contact a law librarian. The local courthouse can instruct you on how to reach a law library if you do not know.

5. Consult with Colleagues Consult with informed colleagues for other ways to consider the problem. This process can be especially helpful in thinking through the circumstances and information and in identifying possible gaps and issues not already considered. Given the responsibility to make a reasoned decision, consulting with colleagues is one way to "act in good faith" and test your justifications. Choose your colleagues wisely.

6. Determine the Goals   One of the most important things to think through is what you hope to see happen as a result of action being taken on the attitudes, behaviors, or circumstances in question. Question your motives carefully and repeatedly. Is your client's voice heard in the goals you want? Talk with a colleague about your goals for a resolution and whether or not there may be motives on your part of which you are not aware. Choose your colleagues wisely.

7. Brainstorm the Strategies   Identify all possible courses of action, including the absurd. Some may prove useful, though unorthodox. Consider the client's perspective as well. Is the client's voice being heard in your list of options? Discuss options with others. Choose your colleagues wisely.

8. Consider the Consequences   Think about the consequences of each strategy for all involved in the situation. Whatever the plans, they must be thoughtfully assessed. Your task is to identify consequences from various perspectives and to question each of the consequences you identify. Remember to include the clients' perspective among those you consider.

9. Consult the Checklist   Use the following checklist as a framework to evaluate potential areas of ethical and legal misconduct. The questions are based in part on a model of ethical decision making that identifies the six fundamental principles of moral behavior: autonomy (self-determination), beneficence (in the best interest of the client), nonmaleficence (to do no harm), justice (fairness to all), fidelity (honest promises and honored commitments), and veracity (being truthful). This model includes such qualities of ethical acts as universality, morality, and reasoned and principled behaviors (Corey et al., 2002; Kitchener, 1984; Pollock, 1998).

- *Is the action in the best interest of the client?* Consider the six fundamental principles of moral behavior.
- *Does the action violate the rights of another person?* Consider constitutional rights as well as your duty to justice.
- *Does the action involve treating another person only as a means to achieve a self-serving end?* Consider the end-in-itself motive and the utilitarian perspective.
- *Is the action under consideration legal? Is it ethical?* Consider the laws and your legal duties; consider your civic and ethical duties, and the components of an ethical act.
- *Does the action create more harm than good for those involved?* Consider the principles of nonmaleficence and beneficence.
- *Does the action violate existing policies, regulations, procedures, or professional standards?* Consider the duty to one's professional role.
- *Does the action promote values in culturally affirming ways?* Consider the principles of nonmaleficence and beneficence and the duty to care.

TABLE 13.2
Ten Steps to Deciding Dilemmas

1. Name the Problem

2. Narrow the Focus

3. Consult the Documents

4. Consider the Laws

5. Consult with Colleagues

6. Determine the Goals

7. Brainstorm the Strategies

8. Consider the Consequences

9. Consult the Checklist

- Principles of autonomy, beneficence, nonmaleficence, justice, fidelity, and veracity

- Duties to care and civic responsibility

- Responsibility to profession's laws, ethical documents, policies, procedures, and regulations

10. Decide with Care

---

10. Decide with Care   Consider carefully the information you have. The more obvious the dilemma, the clearer the course of action; the more nebulous the dilemma, the more difficult the choice. Though hindsight may teach you differently, the best decision in these circumstances is a well-reasoned decision with which you can live.

TAKING CARE OF YOURSELF

One of the potential pitfalls to reading a chapter such as this one is the tendency, regardless of age, to worry unnecessarily about your vulnerability to the issues raised in the reading. You would not be in the field if your academic program did not prepare you adequately for the experience; and, you would not be doing the work you are doing if your site supervisor and field instructor did not believe you were academically and professionally ready for the experience. Even so, issues and situations will present themselves, and they can happen quite quickly. For example, on the first day of a practicum in a courthouse, one of our students was asked to assist in gathering information in interviews resulting from a drug sting. Those arrested were of the same ethnicity and spoke the same first language as did our student. The student was approached by those arrested and was asked to help them evade court processing. Another student was asked out on a date by a convicted offender within the first hour of her practicum at a day reporting center. Both students were shocked at how quickly the circumstances occurred. However, neither of them experienced dilemmas about their situations, though they were surprised to learn how easy it was to find themselves in potentially compromising situations.

Sometimes a situation develops through no fault of your own, and you find yourself at the unenviable end of allegations, complaints, or legal charges. When that happens, it is a very stressful time for all involved, especially for you. You are well aware of what you have invested in your field experience and the importance of your grade and learning to your academic work and/or career plans. However, you may not be aware that such situations do arise and that *how* you handle the situation is very important to your supervisors' assessments of you in crisis.

Knowing how to respond to such a crisis can make all the difference for your future in the profession. Managing this type of crisis is no different than managing other crises you have lived through or half-expect to occur at some time. There are many approaches to managing crises, and it is not our intent to suggest a particular approach to you. You should know what works and what does not work for you. However, we do suggest that you think in terms of a four-prong approach when responding to the crisis.

Self Care in a Crisis

1. Create a crisis resource system.
2. Know yourself.
3. Create a crisis response plan.
4. Learn from the crisis.

Create a Crisis Resource System

Regardless of the approach you use, it is very important that you know the resources that are readily available to you and those that you need to develop. The resources should include but not necessarily be limited to legal, emotional, physical, academic, and professional supports. Knowing beforehand what, who, and where the resources are and how to mobilize them is essential to a healthy and effective response. Otherwise, you can find yourself responding in ways that are not helpful to yourself or the situation. For example, becoming so upset that you do not know which way to turn may be an understandable but not very useful way of responding. Drinking alcohol, taking drugs, or engaging in other self-destructive behaviors is neither professional nor useful to the situation. What is useful, though, is identifying your resources beforehand so you know who to call, where to go, and what you can expect from them. It may be helpful at this point to review the sections in Chapter 4 on support systems and identify supports that you need to develop for yourself.

Know Yourself

It is very important that you understand your reactions, strengths, and weaknesses in a crisis. How you go about solving crises, what works most effectively for you, and what is ineffective in such situations are all informative pieces of information. Thinking these through now, as opposed to when you are in a crisis, is very important because your objectivity is not compromised by the pain of the situation. This also may be a good time to revisit Chapters 3 and 4, especially the parts that help you understand how you function under stress.

Create a Crisis Response Plan

In addition to knowing your resources and knowing yourself in a crisis, it is critical that you have a crisis plan of action, that is, a plan that allows you to be most helpful to yourself and the situation even in times of high anxiety. An important piece of such a plan is a personal crisis team—the people you can call on in an instant to give you the help you need, whether it is legal counsel or chicken soup, literally or figuratively. It is a team of first responders that you create for yourself. In putting together your critical support team, be sure to identify how to reach them when you need them (that is, via e-mail, cell, telephone, or post). Next, you need to think through what you must do to take care of yourself emotionally, physically, and academically throughout the storm so that you stay afloat when the waters get rough. Maybe a physical outlet is best for you, like running, biking, or workouts at the local gym. Perhaps it is yoga that makes a difference in your ability to cope under pressure. Some find prayer or other meditative activities helpful. Many find counseling to be comforting and affirming. Whatever it may be, you need to be aware of it, keep it foremost in your mind, and make it part of your agenda in a crisis.

Learn from the Crisis

Make a resolve to do the best you can under the circumstances and to learn from the crisis in your field experience. It is very important to think through the value of such a resolve now, when you are not in a storm of emotions and when it is easier to appreciate its benefits. Not only does a resolve to learn from a crisis provide you with an understanding of how you function in a crisis within a professional context, it gives you insight into an aspect of the profession that you otherwise would only read about. Taking care of yourself legally, emotionally, physically, academically, and professionally is your responsibility in a crisis. It is also an essential factor in you riding out the storm in ways that leave you the stronger for it.

SUMMARY

You have done a lot of reading about issues in this chapter, most of which are probably new for you. At best, you have become familiar with some terminology and have become aware of areas of concern that you could face in both the field and the profession. This chapter is intended to give you a way of organizing your thinking about the wide variety of issues—professional, ethical and legal—that are part of the profession and about ways to respond should a crisis develop in these aspects of the work. We encourage you to use this chapter as a resource throughout your field experience and to consider taking related academic course work if you have not already done so.

As you continue to develop your sense of competency and professional identity, and deal with your concerns about professional behavior, you are fast approaching the last mile of your journey. We discuss that final mile in the next chapter and the feelings interns experience when their good-byes are bittersweet.

For Further Reflection

FOR PERSONAL REFLECTION:
SELECT THOSE QUESTIONS MOST MEANINGFUL TO YOU

In thinking about the personal system of ethics that you live by, keep in mind that yours might be similar to another person's in some ways, but you will always find areas of differences. Your system of ethics is unique, regardless of the shared ethics that your profession mandates.

1. How do you go about confronting ethical decisions? What words best describe your style: *thoughtful? impulsive? convenient? pressured? expedient?* Others?

2. What values define your personal set of ethics? Your professional set of ethics? Are there areas of potential conflict between the two?

3. Are there legal aspects of your work in the field that are particularly challenging for you to uphold or otherwise live with? Why do you think those aspects are challenging for you?

4. How consistent is your behavior with these values—personally and professionally?

5. Have either your values or ethics changed since you began working in the field of human services? If so, in what ways? What has influenced your ways of thinking? Behaving?

6. What areas of your ethical and/or legal conduct have caused you concern in the past? To what areas do you want to pay particular attention in the future?

7. What are the sources of your systems of values and ethics? How did you come by them? If you could change those circumstances, would you? In what ways? Why?

8. What ethical dilemma do you most remember from childhood? Why do you think that is? How did you deal with it then? Would you deal with it differently now? What life lesson did it give you?

9. Reflect on the following questions, based on an ethical or legal situation or dilemma that you have faced or are facing in your field experience or in a work setting. In what ways did your decisions affect others? Would you make the same decisions if you were on the other side of the issue? How proud are you of your decisions? How proud would your family be?

SPRINGBOARDS FOR DISCUSSION

1. Think about an ethical issue—it does not have to be dilemma or situation—that you would not want to face in your field experience, but one that you could face if the circumstances presented themselves. Share the issues with peers in class, and select one of particular interest to role-play.

2. Think about how you might react to an allegation, complaint, or legal charge brought against you. Would it matter if you were not responsible for the resulting damages? Would it matter if you were responsible and denied it?

3. Discuss with your peers what you would do if you were in a situation where you were accused of wrongdoing and ethical, legal, or professional action was being taken against you. What resources would you activate?

4. Think about the importance of a resolve to learn from a crisis in the field. Discuss with your peers the advantages of making such a resolve and the ways in which it can affect you personally and professionally.

For Further Exploration

Brill, N., & Levine, J. (2001). *Working with people: The helping process* (7th ed.). New York: Longman.

Thoughtful consideration of a variety of professional, moral, and ethical issues in the helping professions.

Corey, G., Corey, M., & Callanan, P. (2003). *Issues and ethics in the helping professions* (6th ed.). Pacific Grove, CA: Brooks/Cole.

Seminal text with a focus on the legal, ethical, or professional dimensions of the helping professions.

Cottone, R. R., & Tarvydas, V. M. (2003). *Ethical and professional issues in counseling* (2nd. ed.). Upper Saddle River, NJ: Merrill/Prentice-Hall.

Comprehensive and informative text focusing on professional, ethical, and legal issues in counseling.

Goldstein, M. B. (1990). Legal issues in combining service and learning. In J. C. Kendall & Associates (Eds.), *Combining service and learning* (Vol. 2, pp. 39–60). Raleigh, NC: National Society for Internships and Experiential Education.

Offers a basic guide to the legal issues involved in combining education and learning in the community.

Kenyon, P. (1999). *What would you do? An ethical case workbook for human service professionals*. Pacific Grove, CA: Brooks/Cole.

Comprehensive and useful workbook that uses actual field situations to inform student awareness of ethical issues and develop the decision-making skills to respond in an ethical manner.

Loewenberg, F., Dolgoff, R., & Harrington, D. (2000). *Ethical decisions for social work practice*. Itasca, IL: F. E. Peacock Publishers.

Comprehensive treatment of the topic and resources for social work students; useful glossary of terms and listing of Internet sources of information about ethics and values.

Rothman, J. C. (2000). *Stepping out into the field: A field work manual for social work students.* Boston: Allyn & Bacon.

Comprehensive guide that provides a useful and thoughtful approach to varied professional issues for social work students.

Steinman, S. O., Richardson, N. F., & McEnroe, T. (1998). *The ethical decision-making manual for helping professionals.* Pacific Grove, CA: Brooks/Cole.

Offers students a pragmatic and focused approach to awareness of ethical issues and responses and developing decision-making skills.

Wadsworth Group. (2003). *Codes of ethics for the helping professions.* Belmont, CA: Wadsworth.

A compilation of the full text of ethical documents for 15 helping professions.

Professional Organizations

The following organizations can provide information about professional standards of practice.

Academy of Criminal Justice Sciences (ACJS)
7319 Hanover Parkway, Suite C
Greenbelt, MD 20770
Telephone: 1-800-757-ACJS
Home page: www.acjs.org

American Association for Marriage and Family Therapy (AAMFT)
112 South Alfred Street
Alexandria, VA 22314
Telephone: 1-703-838-9808
Home page: www.aamft.org

American Counseling Association (ACA)
5999 Stevenson Avenue
Alexandria, VA 22304
Telephone: 1-703-823-9800
Home Page: www.counseling.org

American Psychological Association (APA)
750 First Street, NE, Suite 100
Washington, DC 20003
Telephone: 1-202-336-5500
Home page: www.apa.org

National Association of Social Workers (NASW)
750 First Street, NE, Suite 700
Washington, DC 20002-4241
Telephone: 1-800-638-8799
Home page: www.naswdc.org

National Organization for Human Service Education (NOHSE)
Home page: www.nohse.com

National Society for Experiential Education (NSEE)
9001 Braddock Road, Suite 380
Springfield, VA 22151-1002
Telephone: 1-800-803-4170
Home page: www.nsee.org

References

Alverno College Productions. (1985). *Critical thinking: The Alverno model.* Milwaukee, WI: Author.

American Psychological Association. (1992). *Ethical principles of psychologists and code of conduct.* Washington, DC: Author.

Baird, B. N. (2002). *The internship, practicum, and field placement handbook: A guide for the helping professions.* Upper Saddle River, NJ: Prentice Hall.

Berg-Weger, M., & Birkenmaier, J. (2000). *The practicum companion for social work: Integrating class and field work.* Boston: Allyn & Bacon.

Boylan, J. C., Malley, P. B., & Reilly, E. P. (2001). *Practicum & internship: Textbook and resource guide for counseling and psychotherapy* (3rd ed.). Philadelphia: Bruner-Routledge.

Bradley, L. J., Kottler, J. A., & Lehrman-Waterman, D. (2001). Ethical issues in supervision. In L. J. Bradley & N. Ladany (Eds.), *Counselor supervision: Principles, process and practice* (3rd ed., pp. 342–360). Philadelphia: Bruner-Routledge.

Champion, D. J. (1997). *The Roxbury dictionary of criminal justice.* Los Angeles: Roxbury Publishing Company.

Chiaferi, R., & Griffin, M. (1997). *Developing fieldwork skills.* Pacific Grove, CA: Brooks/Cole.

Close, D., & Meier, N. (1995). *Morality in criminal justice.* Belmont, CA: Wadsworth.

Collins, D., Thomlison, B., & Grinnell, R. M. (1992). *The social work practicum: A student guide.* Itasca, IL: F. E. Peacock.

Corey, G., Corey, M., & Callanan, P. (2003). *Issues and ethics in the helping professions* (6th ed.). Pacific Grove, CA: Brooks/Cole.

Corey, M. S., & Corey, G. (2003). *Becoming a helper* (4th ed.). Pacific Grove, CA: Brooks/Cole.

Dougherty, A. M. (1995). *Consultation: Practice and perspective in school and community settings* (2nd ed.). Pacific Grove, CA: Brooks/Cole.

Faiver, C., Eisengard, S., & Colonna, R. (2000). *The counselor intern's handbook* (2nd ed.). Belmont, CA: Wadsworth.

Falvey, J., with Bray, T. (2002). *Managing clinical supervision: Ethical practice and legal risk management.* Pacific Grove, CA: Brooks/Cole.

Gibbs, L., & Gambrill, E. (1996). *Critical thinking for social workers: A workbook.* Thousand Oaks, CA: Pine Forge Press.

Goldstein, M. B. (Ed.). (1990). Legal issues in combining service and learning. In J. C. Kendall & Associates (Eds.), *Combining service and learning* (Vol. 2, pp. 39–60). National Society for Internships and Experiential Education.

Gordon, G., McBride, R. B., & Hage, H. H. (2001). *Criminal justice internships: Theory into practice* (4th ed.). Cincinnati, OH: Anderson Publishing.

Horejsi, C., & Garthwait, C. (2002). *The social work practicum: A guide and workbook for students.* Needham Heights, MA: Allyn & Bacon.

Kanter, R. M. (1977). *Men and women of the corporation.* New York: Basic Books.

Kenyon, P. (1999). *What would you do?: An ethical case workbook for human service professionals.* Pacific Grove, CA: Brooks/Cole.

Kiser, P.M. (2000). *Getting the most from your human service internship.* Belmont, CA: Brooks/Cole.

Kitchener, K. S. (1984). Intuition, critical evaluation and ethical principles: The foundation for ethical decisions in counseling psychology. *Counseling Psychologist, 12*(3), 43–45.

Malley, P., & Reilly, E. (2001). Legal issues. In J. Boylan, P. Malley, & E. Reilly (Eds.), *Practicum & internship: Textbook and resource guide for counseling and psychotherapy* (3rd ed., pp. 129–142). Philadelphia: Brunner-Routledge.

Martin, M. L. (Ed.). (1991). *Employment setting as practicum site: A field instruction dilemma.* Dubuque, IA: Kendall/Hunt.

Meara, N.M., Schmidt, L. D., & Day, J. D. (1996). Principles and virtues: A foundation for ethical decisions, policies and character. *Counseling Psychologist, 24*(1), 4–77.

National Organization for Human Service Education. (2000). Ethical standards of human service professionals. *Human Service Education, 20*(1), 61–68.

National Society for Experiential Education. (1998). *Standards of practice: Eight principles of good practice for all experiential learning activities.* Presented at the Annual Meeting, Norfolk, VA. Washington, DC: Author.

Neukrug, E. (2000). *Theory, practice and trends in human services: An overview of an emerging profession* (2nd ed.). Pacific Grove, CA: Brooks/Cole.

Oran, D. (1985). *Law dictionary for nonlawyers* (2nd ed.). St. Paul, MN: West Publishing Co.

Pollock, J. M. (2004). *Ethics in crime and justice: Dilemmas and decisions* (4th ed.). Belmont, CA: Wadsworth.

Rothman, J. C. (2000). *Stepping out into the field: A field work manual for social work students.* Boston: Allyn & Bacon.

Royse, D., Dhooper, S. S., & Rompf, E. L. (2003). *Field instruction: A guide for social work students* (4th ed.). New York: Longman.

Schram, B., & Mandell, B. R. (2002). *An introduction to human services: Policy and practice* (5th ed.). New York: Macmillan.

Schultz, M. (1992). Internships in sociology: Liability issues and risk management measures. *Teaching Sociology, 20*, 183–191.

Steinman, S. O., Richardson, N. F., & McEnroe, T. (1998). *The ethical decision-making manual for helping professionals.* Pacific Grove, CA: Brooks/Cole.

Tarvydas, V., Cottone, R., & Claus, R. (2003). Ethical decision-making processes. In R. Cottone & V. Tarvydas (Eds.), *Ethical and professional issues in counseling* (2nd ed.). Upper Saddle River, NJ: Merrill/Prentice-Hall.

Taylor, D. (1999). *Jumpstarting your career: An internship guide for criminal justice.* Upper Saddle River, NJ: Prentice Hall.

Tentoni, S. C. (1995). The mentoring of counseling students: A concept in search of a paradigm. *Counselor Education and Supervision, 35*(1), 32–41.

Travers, P. (2002). *The counselor's helpdesk.* Pacific Grove, CA: Brooks/Cole.

Tryon, G. S. (1996). Supervisee development during the practicum year. *Counselor Education and Supervision, 35*(4), 287–294.

Wadsworth Group. (2003). *Codes of ethics for the helping professions.* Belmont, CA: Wadsworth.

Wilson, S. J. (1981). *Field instruction: Techniques for supervisors.* New York: Free Press.

Woodside, M., & McClam, T. (2002). *An introduction to human services* (4th ed.). Pacific Grove, CA: Brooks/Cole.

APPENDIX

Standards of Practice:
Eight Principles of Good Practice
for All Experiential Learning Activities

NATIONAL SOCIETY FOR EXPERIENTIAL LEARNING

Regardless of the experiential learning activity, both the experience and the learning are fundamental. In the learning process and in the relationship between the learner and any facilitator(s) of learning, there is a mutual responsibility. All parties are empowered to achieve the principles which follow. Yet, at the same time, the facilitator(s) of learning are expected to take the lead in ensuring both the quality of the learning experience and of the work produced, and in supporting the learner to use the principles which underlie the pedagogy of experiential education.

1. Intention: All parties must be clear from the outset why experience is the chosen approach to the learning that is to take place and to the knowledge that will be demonstrated, applied or result from it. Intention represents the purposefulness that enables experience to become knowledge and, as such, is deeper than the goals, objectives, and activities that define the experience.

2. Preparedness and Planning: Participants must ensure that they enter the experience with sufficient foundation to support a successful experience. They must also focus from the earliest stages of the experience/program on the identified intentions, adhering to them as goals, objectives and activities are defined. The resulting plan should include those intentions and be referred to on a regular basis by all parties. At the same time, it should be flexible enough to allow for adaptations as the experience unfolds.

3. Authenticity: The experience must have a real world context and/or be useful and meaningful in reference to an applied setting or situation. This means that is should be designed in concert with those who will be affected by or use it, or in response to a real situation.

4. Reflection: Reflection is the element that transforms simple experience to a learning experience. For knowledge to be discovered and internalized, the learner must test assumptions and hypotheses about the outcomes of decisions and actions taken, then weigh the outcomes against past learning and future implications. This reflective process is integral to all phases of experiential learning, from identifying intention and choosing the experience, to considering preconceptions and observing how they

Source: National Society for Experiential Education. (1998). *Standards of Practice: Eight Principles of Good Practice for All Experiential Learning Activities.* Presented at the Annual Meeting, Norfolk, VA. Washington, DC: Author. Reprinted with permission.

change as the experience unfolds. Reflection is also an essential tool for adjusting the experience and measuring outcomes.

5. Orientation and Training: For the full value of the experience to be accessible to both the learner and the learning facilitator(s), and to any involved organizational partners, it is essential that they be prepared with important background information about each other and about the context and environment in which the experience will operate. Once that baseline of knowledge is addressed, ongoing structured development opportunities should also be included to expand the learner's appreciation of the context and skill requirements of her/his work.

6. Monitoring and Continuous Improvement: Any learning activity will be dynamic and changing, and the parties involved all bear responsibility for ensuring that the experience, as it is in process, continues to provide the richest learning possible, while affirming the learner. It is important that there be a feedback loop related to learning intentions and quality objectives and that the structure of the experience be sufficiently flexible to permit change in response to what that feedback suggests. While reflection provides input for new hypotheses and knowledge based in documented experience, other strategies for observing progress against intentions and objectives should also be in place. Monitoring and continuous improvement represent the formative evaluation tools.

7. Assessment and Evaluation: Outcomes and processes should be systematically documented with regard to initial intentions and quality outcomes. Assessment is a means to develop and refine the specific learning goals and quality objectives identified during the planning stages of the experience, while evaluation provides comprehensive data about the experiential process as a whole and whether it has met the intentions which suggested it.

8. Acknowledgment: Recognition of learning and impact occur throughout the experience by way of the reflective and monitoring processes and through reporting, documentation and sharing of accomplishments. All parties to the experience should be included in the recognition of progress and accomplishment. Culminating documentation and celebration of learning and impact help provide closure and sustainability to the experience.

Ethical Standards
of Human Service Professionals

NATIONAL ORGANIZATION FOR HUMAN SERVICE EDUCATION
COUNCIL FOR STANDARDS IN HUMAN SERVICE EDUCATION

Preamble

Human services is a profession developing in response to and in anticipation of the direction of human needs and human problems in the late twentieth century. Characterized particularly by an appreciation of human beings in all of their diversity, human services offers assistance to its clients within the context of their community and environment. Human service professionals, regardless of whether they are students, faculty or practitioners, promote and encourage the unique values and characteristics of human services. In so doing human service professionals uphold the integrity and ethics of the profession, partake in constructive criticism of the profession, promote client and community well-being, and enhance their own professional growth.

The ethical guidelines presented are a set of standards of conduct which the human service professional considers in ethical and professional decision making. It is hoped that these guidelines will be of assistance when the human service professional is challenged by difficult ethical dilemmas. Although ethical codes are not legal documents, they may be used to assist in the adjudication of issues related to ethical human service behavior.

Human service professionals function in many ways and carry out many roles. They enter into professional-client relationships with individuals, families, groups and communities who are all referred to as "clients" in these standards. Among their roles are caregiver, case manager, broker, teacher/educator, behavior changer, consultant, outreach professional, mobilizer, advocate, community planner, community change organizer, evaluator and administrator. The following standards are written with these multifaceted roles in mind.

The Human Service Professional's Responsibility to Clients

STATEMENT 1 Human service professionals negotiate with clients the purpose, goals, and nature of the helping relationship prior to its onset as well as inform clients of the limitations of the proposed relationship.

STATEMENT 2 Human service professionals respect the integrity and welfare of the client at all times. Each client is treated with respect, acceptance and dignity.

STATEMENT 3 Human service professionals protect the client's right to privacy and confidentiality except when such confidentiality would cause harm to the client or

Source: From *Human Service Education*, 16(1), by the National Organization for Human Service Education, pp. 11–17. Reprinted with permission.

others, when agency guidelines state otherwise, or under other stated conditions (e.g., local, state, or federal laws). Professionals inform clients of the limits of confidentiality prior to the onset of the helping relationship.

STATEMENT 4 If it is suspected that danger or harm may occur to the client or to others as a result of a client's behavior, the human service professional acts in an appropriate and professional manner to protect the safety of those individuals. This may involve seeking consultation, supervision, and/or breaking the confidentiality of the relationship.

STATEMENT 5 Human service professionals protect the integrity, safety, and security of client records. All written client information that is shared with other professionals, except in the course of professional supervision, must have the client's prior written consent.

STATEMENT 6 Human service professionals are aware that in their relationships with clients, power and status are unequal. Therefore they recognize that dual or multiple relationships may increase the risk of harm to, or exploitation of, clients, and may impair their professional judgment. However, in some communities and situations it may not be feasible to avoid social or other nonprofessional contact with clients. Human service professionals support the trust implicit in the helping relationship by avoiding dual relationships that may impair professional judgment, increase the risk of harm to clients or lead to exploitation.

STATEMENT 7 Sexual relationships with current clients are not considered to be in the best interest of the client and are prohibited. Sexual relationships with previous clients are considered dual relationships and are addressed in Statement 6 (above).

STATEMENT 8 The client's right to self-determination is protected by human service professionals. They recognize the client's right to receive or refuse services.

STATEMENT 9 Human service professionals recognize and build on client strengths.

The Human Service Professional's Responsibility to the Community and Society

STATEMENT 10 Human service professionals are aware of local, state, and federal laws. They advocate for change in regulations and statutes when such legislation conflicts with ethical guidelines and/or client rights. Where laws are harmful to individuals, groups or communities, human service professionals consider the conflict between the values of obeying the law and the values of serving people and may decide to initiate social action.

STATEMENT 11 Human service professionals keep informed about current social issues as they affect the client and the community. They share that information with clients, groups and community as part of their work.

STATEMENT 12 Human service professionals understand the complex interaction between individuals, their families, the communities in which they live, and society.

STATEMENT 13 Human service professionals act as advocates in addressing unmet client and community needs. Human service professionals provide a mechanism for identifying unmet client needs, calling attention to these needs, and assisting in planning and mobilizing to advocate for those needs at the local community level.

STATEMENT 14 Human service professionals represent their qualifications to the public accurately.

STATEMENT 15 Human service professionals describe the effectiveness of programs, treatments, and/or techniques accurately.

STATEMENT 16 Human service professionals advocate for the rights of all members of society, particularly those who are members of minorities and groups at which discriminatory practices have historically been directed.

STATEMENT 17 Human service professionals provide services without discrimination or preference based on age, ethnicity, culture, race, disability, gender, religion, sexual orientation or socioeconomic status.

STATEMENT 18 Human service professionals are knowledgeable about the cultures and communities within which they practice. They are aware of multiculturalism in society and its impact on the community as well as individuals within the community. They respect individuals and groups, their cultures and beliefs.

STATEMENT 19 Human service professionals are aware of their own cultural backgrounds, beliefs, and values, recognizing the potential for impact on their relationships with others.

STATEMENT 20 Human service professionals are aware of sociopolitical issues that differentially affect clients from diverse backgrounds.

STATEMENT 21 Human service professionals seek the training, experience, education and supervision necessary to ensure their effectiveness in working with culturally diverse client populations.

The Human Service Professional's Responsibility to Colleagues

STATEMENT 22 Human service professionals avoid duplicating another professional's helping relationship with a client. They consult with other professionals who are assisting the client in a different type of relationship when it is in the best interest of the client to do so.

STATEMENT 23 When a human service professional has a conflict with a colleague, he or she first seeks out the colleague in an attempt to manage the problem. If necessary, the professional then seeks the assistance of supervisors, consultants or other professionals in efforts to manage the problem.

STATEMENT 24 Human service professionals respond appropriately to unethical behavior of colleagues. Usually this means initially talking directly with the colleague and, if no resolution is forthcoming, reporting the colleague's behavior to supervisory or administrative staff and/or to the professional organization(s) to which the colleague belongs.

STATEMENT 25 All consultations between human service professionals are kept confidential unless to do so would result in harm to clients or communities.

The Human Service Professional's
Responsibility to the Profession

STATEMENT 26 Human service professionals know the limit and scope of their professional knowledge and offer services only within their knowledge and skill base.

STATEMENT 27 Human service professionals seek appropriate consultation and supervision to assist in decision-making when there are legal, ethical or other dilemmas.

STATEMENT 28 Human service professionals act with integrity, honesty, genuineness, and objectivity.

STATEMENT 29 Human service professionals promote cooperation among related disciplines (e.g., psychology, counseling, social work, nursing, family and consumer sciences, medicine, education) to foster professional growth and interests within the various fields.

STATEMENT 30 Human service professionals promote the continuing development of their profession. They encourage membership in professional associations, support research endeavors, foster educational advancement, advocate for appropriate legislative actions, and participate in other related professional activities.

STATEMENT 31 Human service professionals continually seek out new and effective approaches to enhance their professional abilities.

The Human Service Professional's
Responsibility to Employers

STATEMENT 32 Human service professionals adhere to commitments made to their employers.

STATEMENT 33 Human service professionals participate in efforts to establish and maintain employment conditions which are conducive to high quality client services. They assist in evaluating the effectiveness of the agency through reliable and valid assessment measures.

STATEMENT 34 When a conflict arises between fulfilling the responsibility to the employer and the responsibility to the client, human service professionals advise both of the conflict and work conjointly with all involved to manage the conflict.

The Human Service Professional's
Responsibility to Self

STATEMENT 35 Human service professionals strive to personify those characteristics typically associated with the profession (e.g., accountability, respect for others, genuineness, empathy, pragmatism).

STATEMENT 36 Human service professionals foster self-awareness and personal growth in themselves. They recognize that when professionals are aware of their own values, attitudes, cultural background, and personal needs, the process of helping others is less likely to be negatively impacted by those factors.

STATEMENT 37 Human service professionals recognize a commitment to lifelong learning and continually upgrade knowledge and skills to serve the populations better.

Section II: Standards for Human Service Educators

Human Service educators are familiar with, informed by and accountable to the standards of professional conduct put forth by their institutions of higher learning; their professional disciplines, for example, the American Association of University Professors (AAUP), American Counseling Association (ACA), Academy of Criminal Justice Society (ACJS), American Psychological Association (APA), American Sociological Association (ASA), National Association of Social Workers (NASW), National Board of Certified Counselors (NBCC), National Education Association (NEA); and the National Organization for Human Service Education (NOHSE).

STATEMENT 38 Human service educators uphold the principle of liberal education and embrace the essence of academic freedom, abstaining from inflicting their own personal views/morals on students, and allowing students the freedom to express their views without penalty, censure or ridicule, and to engage in critical thinking.

STATEMENT 39 Human service educators provide students with readily available and explicit program policies and criteria regarding program goals and objectives, recruitment, admission, course requirements, evaluations, retention and dismissal in accordance with due process procedures.

STATEMENT 40 Human service educators demonstrate high standards of scholarship in content areas and of pedagogy by staying current with developments in the field of Human Services and in teaching effectiveness, for example learning styles and teaching styles.

STATEMENT 41 Human service educators monitor students' field experiences to ensure the quality of the placement site, supervisory experience, and learning experience towards the goals of professional identity and skill development.

STATEMENT 42 Human service educators participate actively in the selection of required readings and use them with care, based strictly on the merits of the material's content, and present relevant information accurately, objectively and fully.

STATEMENT 43 Human service educators, at the onset of courses: inform students if sensitive/controversial issues or experiential/affective content or process are part of the course design; ensure that students are offered opportunities to discuss in structured ways their reactions to sensitive or controversial class content; ensure that the presentation of such material is justified on pedagogical grounds directly related to the course; and, differentiate between information based on scientific data, anecdotal data, and personal opinion.

STATEMENT 44 Human service educators develop and demonstrate culturally sensitive knowledge, awareness, and teaching methodology.

STATEMENT 45 Human service educators demonstrate full commitment to their appointed responsibilities, and are enthusiastic about and encouraging of students' learning.

STATEMENT 46 Human service educators model the personal attributes, values and skills of the human service professional, including but not limited to, the willingness to seek and respond to feedback from students.

STATEMENT 47 Human service educators establish and uphold appropriate guidelines concerning self-disclosure or student-disclosure of sensitive/personal information.

STATEMENT 48 Human service educators establish an appropriate and timely process for providing clear and objective feedback to students about their performance on relevant and established course/program academic and personal competence requirements and their suitability for the field.

STATEMENT 49 Human service educators are aware that in their relationships with students, power and status are unequal; therefore, human service educators are responsible to clearly define and maintain ethical and professional relationships with students, and avoid conduct that is demeaning, embarrassing or exploitative of students, and to treat students fairly, equally and without discrimination.

STATEMENT 50 Human service educators recognize and acknowledge the contributions of students to their work, for example in case material, workshops, research, publications.

STATEMENT 51 Human service educators demonstrate professional standards of conduct in managing personal or professional differences with colleagues, for example, not disclosing such differences and/or affirming a student's negative opinion of a faculty/program.

STATEMENT 52 Human service educators ensure that students are familiar with, informed by, and accountable to the ethical standards and policies put forth by their program/department, the course syllabus/instructor, their advisor(s), and the Ethical Standards of Human Service Professionals.

STATEMENT 53 Human service educators are aware of all relevant curriculum standards, including those of the Council for Standards in Human Services Education (CSHSE); the Community Support Skills Standards; and state/local standards, and take them into consideration in designing the curriculum.

STATEMENT 54 Human service educators create a learning context in which students can achieve the knowledge, skills, values and attitudes of the academic program.

CHAPTER *14*

Traveling the Last Mile: The Culmination Stage

I've been taking in as much information as I can before "it's all over" and have been concerned with career goals. The past couple of weeks have just been filled with an overwhelming amount of anxieties and mixed emotions.

STUDENT REFLECTION

As incredible as it must seem to you at times, the end is in sight. For most of you, it is the end of something special, although for some of you, it may not have been the experience you hoped for. In either case, the beginning of your internship may seem like yesterday; or it may seem like years ago; or you may go back and forth between those two feelings. It is a time to look forward to the future and to be proud of what you have accomplished. So why don't you always feel clear about the experience and about ending it?

In our experience, interns approaching the end of their field experience report many different feelings. There is often pride and sometimes feelings of mastery. Others report relief and anticipation of freedom. A different kind of relief is expressed by interns who have struggled or had a disappointing experience. However, we also hear about sadness, anger, loss, and confusion, regardless of how satisfying the internship has been. In a survey of psychotherapy interns, Robert Gould (1978) found that many of them reported increased anxiety and depression as the end approached as well as decreased effectiveness. The interns in his study also reported feeling moodier. That has been our experience as well; you would not be unusual if you found yourself experiencing several—or even all—of these feelings, simultaneously or sequentially, with your emotional landscape shifting by the hour. What is going on here?

For one thing, you have a lot on your plate right now. Your internship is ending, and you may or may not be ready. The calendar says the internship is over, but you

263

may feel as though you have just gotten started, especially if it took some time to hit your stride (Suelzle & Borzak, 1981). In addition, you may or may not have completed the work you set out for yourself. Projects may not be finished; clients may not be quite where you would like them to be. On the other hand, you may feel as though you have been finished for a while.

The web of relationships that has been the social context of your internship is changing yet again. Many relationships are ending; others will be redefined as you leave the role of intern. Endings are part of most relationships and a necessary part of helping relationships (Brill & Levine, 2001). The more the relationships mean to you, though, the harder it will be to end them. Some interns find that the impact of relationships actually reaches its peak as the relationships near their end (Gould, 1978):

> Now I am really sad. I loved my internship and don't want to lose the relationships I've formed here. I'm trying to make plans with everyone so we can keep in touch.

The external context of your life is shifting as well. There may be new demands on your time and energy as you feel the pressure of endings and beginnings (Suelzle & Borzak, 1981). Your attention may be pulled toward papers and final projects for your internship seminar as well as to papers and exams in other classes. Your summer or holiday plans may need to be finalized. If you are graduating and have not yet found a job, your job search may be occupying more and more time and attention. Of course, your internship is not the only thing that is ending. Your seminar class, the semester, the school year, even your college career may all be drawing to a close. That's a lot of endings, a lot of beginnings, and a lot of good-byes. Good-byes are never easy, and for some people they can be very stressful. The prospect of beginning in a new school, a new town, a new job, or all of the above is daunting as well as exciting.

That is why we call this the Culmination stage; everything is coming to a crescendo. Another stage, another crisis, and another set of risks and opportunities await you. Like other critical junctures discussed in this book, this one is normal, and so are the concerns and issues associated with it. Endings are a necessary part of anyone's development (Kegan, 1982). Normal does not mean easy, though. Remember the distinction between difficulties and problems discussed in Chapter 11? You could think of endings as difficulties that can either be solved, resolved, or aggravated into problems (Watzlawick, Weaklund, & Fisch, 1974).

There are opportunities here for you to grow, to develop new insights, and to learn more about separation and moving on, but there are also risks. It is natural to want to avoid the conflicting feelings that can surface at this time, and that avoidance can take many forms, including lateness or absence, devaluing the experience (suddenly it doesn't seem all that great anymore), or putting on rose-colored glasses and forgetting all the struggles you had and may still be having. There is even a name for these behaviors: "premature disengagement" (Kiser, 2000). The risk in these behaviors is that you can turn a difficulty into a problem. Lateness and absence will not help anything and will create new problems for you. Becoming out of touch with your own emotions may make you less available to clients. You also risk losing the opportunity to learn how to face the issues and feelings associated with endings. Finally, interns who do not face these issues often report a hollow feeling as they leave their placements, even though they may have been good experiences in many ways.

SEIZING THE OPPORTUNITIES

The rest of this chapter is designed to help you minimize the risks, meet the challenges, and maximize the opportunities. Making time to process your experience, taking a new self-inventory, and facing the tasks of ending are all essential components of this effort.

As your internship ends, you may feel more than ever the pull to do more and more. There may be a lot that can or needs to be finished, but becoming too involved in a frenzy of activity can also be a way to avoid facing your feelings. Once again, we remind you—and it is especially important now—that you need time and energy for reflection. You will have to protect that time more jealously than ever, as the external pressures mount and the temptation to avoid your feelings grows stronger.

Taking a Self-Inventory

As you head into this phase of your internship, take some time to think about yourself. All these endings and beginnings are likely to tap a variety of emotional issues. Some of the issues you thought about in Chapters 3 and 4 may resurface or surface for the first time. You may also discover some new issues. Don't let them catch you by surprise.

Although this may be your first internship, don't be so sure the experience of culminating is entirely new for you. You have had endings, separations, trauma, and loss before, and these experiences can leave people with some unfinished emotional business. Your experience of ending the internship will be colored by those experiences, as well as by your response patterns, your psychosocial issues, and even your family patterns.

Think back on the experiences of separation and loss in your life. Going away to camp or to college, having close siblings leave home, moving, divorce, ending an intimate relationship, and being fired from a job are examples of separation experiences you may have had. Those experiences are never easy, and for some people in some circumstances, they can be traumatic. Perhaps some of the hurts from those experiences have not healed. Perhaps you wish you had behaved differently or that others had.

Now think about how you say good-bye and how people have said good-bye to you. Some people just leave and don't say a word; others write long letters or schedule good-bye lunches or dinners. The way you say good-bye probably depends somewhat on the nature of the relationship that is ending, but perhaps you can discern some patterns in how you approach this task. Remember the discussion of dysfunctional patterns in Chapter 3? This is an area in which many people have those patterns, and they can surface in an internship. Here is one we have seen many times:

> Saying good-bye to someone I don't care that much about is pretty easy. When it's someone I'm really invested in, though, I don't handle it well. Usually what I do is get really busy. I keep promising myself to go to lunch, or dinner, or something with the person, but I never seem to make the time. Then all of a sudden there is no time, and I end up saying a hurried good-bye. I know I have hurt people's feelings that way.

Scott Haas (1990), in a book about his psychology internship, notices himself falling into a similar pattern as the end of his internship approaches:

I create obstacles to avoid thinking about the end. I distract myself: I make lists. I ruminate about minor inconveniences (like wondering for days whether the gas company will correct their bill). I develop new projects and interests. I go shopping, and then in the store can't remember why I ever wanted the thing I'm about to buy. When all else fails, I pretend there is no end. I'm just imagining it: it really isn't happening. (p. 171)

Another area that may be touched by the culmination stage is your feelings about the issues of competence. You may recall from Chapter 3 that this is one of the components of your psychosocial identity. Some interns report that they feel competent for the very first time during their internship, and that is very hard to leave behind. Others report feeling less competent at the end than the beginning, even though that is almost certainly not true.

Remember the relationship of competence to success. If your sense of competence is tied to success, then you may be trying to get in one more success before you leave, pursuing perfection in a project or one more milestone with a client. Richard Schafer (1973) believes that many helping professionals strive for a "sense of goodness." Although achievement is part of this sense, another component is feeling like a good person, regardless of one's faults. Gould (1978) points out that this particular sense of goodness can be difficult for an intern to achieve. Many people entering the helping professions, he says, want and need to be liked and seek approval from their clients. Their sense of competence, and goodness, is tied to client success or to some other form of success at their site.

On the other hand, all that some of you can think about is how you did not succeed, or the ways in which you did not succeed. If you had some substantial struggles, if you found out that the sort of work done at your internship is not for you after all, those are hard lessons. But they are also good, important lessons, and they do not make you incompetent.

Being aware of how you deal with endings and knowing where you are vulnerable will help you be open and sensitive to the experience of your clients. You can avoid blaming others for your feelings and experiences. Finally, you can make conscious efforts to avoid the pitfalls.

The Tasks of Ending

According to Naomi Brill and Joanne Levine (2001), there are three tasks associated with what we are calling the Culmination stage. They are not necessarily completed in sequential order; in fact, you will surely see that they are interconnected and are usually dealt with simultaneously. One important task is that *unfinished business must be identified and dealt with*. These are issues with clients, supervisors, coworkers, and yourself that have been present for some time but that often take on added urgency as the internship draws to a close (Shulman, 1983). Next, *you must identify your feelings and find a safe place to express them*. You may find strong feelings about supervisors, clients, peers, and others coming to the surface. It may or may not be appropriate to express these feelings directly to the person who has engendered them. At the very least, you will want to find a place where you can express them to someone else and say whatever you want to say before worrying about just what to say to the people involved. Finally, *you need to plan for the future.*

These three tasks can be applied to the work you are doing, the people you are interacting with, and the placement site as a whole. In some cases, the tasks can be attended to informally, and each intern will do them differently. In other cases, however, we are going to suggest that you be formal and structured about it.

Some of these tasks will be taken care of almost by themselves. They will happen naturally, or someone else will initiate them. Others will not, though, and in those cases you will need to be more proactive. In writing about ending an internship, some authors have discussed the need for rituals (Baird, 2002; Kiser, 2000; Rogers, Collins, Barlow, & Grinnell, 2000). That may sound like a strange term to you, even calling up religious or pagan scenes. However, any formal way of marking an event or passage (such as a going-away luncheon) can be considered a ritual. In this case, rituals can help provide a sense of completion or closure. Brian Baird (2002), whose writings about the end of an internship are especially thoughtful, discusses several benefits of rituals in the closing phase. Rituals can add a sense of specialness to the ending; they can help you recall what was significant and important. They can also help ease the transition by connecting the past to the future. Finally, rituals can create a formal, structured opportunity to experience and express emotions that might otherwise be repressed.

FINISHING THE WORK

As you enter the last weeks of your internship, you need to think about what tasks remain and what you want to and can do about them. The nature of the tasks makes a difference here. If you have been given a series of small, concrete projects, such as a report to write or an event to plan, and they are complete or near completion, then ending the internship will be somewhat easier (Suelzle & Borzak, 1981). If you are part of a large project that will not end until after you have gone, then feeling some closure about your involvement can be harder. Perhaps there is a component of the project you can complete or a summary of your work that you can write. Finally, you may be involved in a complex project that must be completed before you leave. Perhaps, for example, you have done some research for your supervisor and have agreed to summarize your findings. There may be the temptation to read one more article or interview one more person; in some cases, there is always more research you can do. At a certain point, however, you have to stop generating and start summarizing.

Planning for the future also means considering the nature of your involvement, if any, with the agency and its work after your internship is over. Some interns are offered part-time work, relief work, or even full-time jobs. That is great when it happens; just make sure that accepting is really your best option and not just a way to avoid bringing things to an end. In other cases, you may want to come back for a visit or to see and help with an event you have been working on. Coming back for a visit can be fine depending on the nature of your clients (this issue will be discussed more in a later part of the chapter).

Seeing a project or event you have helped plan come to fruition can be wonderful as well. Just take care not to promise more than you can deliver. If the agency is counting on you and you get caught up in your life and don't follow through, you leave a bad feeling about you and possibly about your school.

SAYING GOOD-BYE TO CLIENTS

"Just when I'm really starting to feel comfortable with the kids it's almost time to end," wrote one intern. *Termination* is the word used to describe the ending of a therapeutic relationship. To some of you, this is a familiar term. You may have studied it in class, and if your agency does one-to-one counseling, you may have heard the term there as well. For others the term may be new, somewhat strange, and harsh sounding, especially given its recent use in the movies and on television. Not all of you are doing counseling or psychotherapy in your internship, but if you are doing any sort of direct work with clients, you need to think carefully about ending your work with them and saying good-bye, regardless of the label you apply to the process.

Furthermore, as an intern you are saying good-bye under unusual circumstances. Usually when a client-worker relationship ends, it is because the work is done or the client decides to end it. However, you are dealing with what are called *forced terminations* (Baird, 2002; Gould, 1978). It is the calendar that dictates the ending. The work is not necessarily over, and the clients are not necessarily ready to stop or start over with someone else. Ending relationships with clients is bound to engender emotional reactions for them and for you. Dealing with those feelings and ending well is an important part of the internship. For some of you, it will be relatively easy. For others, it may be one of the biggest challenges you face.

Part of the reason our interns have such varied experiences in coming to closure with clients is that there are such great variations in the internship sites. Some sites do a lot of one-to-one counseling. Others, such as adolescent shelters, do some, but in the context of daily living and recreation. Other sites do none at all. There is also great variation in the characteristics of the clients. Some client populations are much more emotionally vulnerable and may take your departure that much harder. Some clients are more autonomous than others. Clients who drop into a senior services center, for example, or a town recreation department are probably functioning pretty well on their own. They appreciate you, but they really don't need you. Finally, some clients are simply more aware of your leaving than others. One intern, who was working with a number of clients who had Alzheimer's disease, reported that she had to tell each client over and over that she would be leaving. Many of them did not remember her from day to day, although they seemed glad to see her. Leaving was difficult for her, but probably not for them.

In any case, there are four important issues to think about regarding closure with your clients: (a) you need to decide when and how to tell them, (b) you need to identify and take care of any unfinished business with them, (c) you need to deal with your feelings and theirs about your leaving, and (d) you need to plan for their future needs.

Timing and Style

There are many different theories and opinions about when and how to tell clients you are leaving. Because internship sites and client populations vary so greatly, you must look to your site, your supervisor, and your instructor to guide you in this area; no prescriptions we could make would apply to every intern, or even to most of them. Some agencies recommend that you tell clients right away that you will be leaving at the end of the term or the year; in fact, some agencies tell the clients for you. Some clients are used to seeing interns come and go; as soon as you say you are an intern, they know you will not be there for very long. Other agencies recommend that you begin discussion of termination one week, two weeks, or more in advance. Some recommend that you not mention it until right before you go.

There are also a variety of methods of discussing termination, from individual conferences, to group meetings, to letters, or any combination of these approaches. If you do not know how your agency handles termination, you need to take the initiative to ask. Depending on your needs and attitudes about ending, you may feel that the agency's policy is too casual or forces you to focus on something that does not seem like a big deal. Remember that the clients are the primary focus; the decision needs to be made based on what will work best for them, not for you. Still, this is a good opportunity for discussion with your supervisor if you disagree with the agency's policy, or if it does not match a theory you have studied. Both of you might learn something, and at the very least you will be more likely to accept and follow through with the agency's procedures.

Unfinished Business

As you approach the end of your relationship with clients, it is a good time to reflect, certainly in private and perhaps with clients, on their progress. If they had specific goals, this is the time to review them. Even if they did not, your memory and your journal are good sources for remembering what your clients were like when you first arrived, as well as what your relationship with them was like. Both clients and workers can become so caught up in the problems and challenges of the present that they don't think much about the change that has already occurred. Add to that the difficult feelings that termination can bring, and you can see why it is important to take time to reflect on the positive.

It is equally important, though, to be clear with yourself and with clients about the work that remains. All of this is especially important if the client is going to continue at the agency, which is usually the case. You may want to use some of your supervision time to review each of your clients with your supervisor, or you may want to discuss them as a group. You will want to talk about how you feel about each of these terminations, as well as how you think each client is going to react. Baird discusses a Termination Scale developed by Fair and Bressler that has 55 items covering emotional responses and planning. You may want to make use of this resource (Fair & Bressler, cited in Baird, 2002).

Dealing with Feelings

In many cases, the termination process can be an emotional one for everyone involved. Clients may be especially vulnerable. For them, just as for you, your leaving may recall echoes of their past experiences of separation and loss. For many of these clients—

especially those who have been subjected to abuse, neglect, or parents with substance abuse problems—separations have been arbitrary and unpredictable, such as when a parent abandons a child or an alcoholic parent goes on a drinking binge and returns days later. For some people, separation is associated with anger and hysteria, as in the loud and even violent arguments that can lead to the disruption or ending of a marriage or other intimate relationship. You may also find that clients have had particularly traumatic separations, as in the death of a parent, sibling, or close friend. Finally, many of your clients will have worked with several agencies, workers, or several residential settings and have had many termination experiences. If some or all of these experiences have been painful, this termination may bring up those feelings again (Baird, 2002; Kiser, 2000), and you are right to be concerned.

> *"Why do you have to go?" is probably the most difficult [question]. Many of these children have had people who come and go in their lives and I don't want them to look at me as one of those people.*
>   STUDENT REFLECTION

Of course, termination is also a learning opportunity. Clients can learn that good-byes, while often painful, do not have to be traumatic, angry, or hysterical.

Many interns report feeling nervous and apprehensive about discussing termination with clients because they are afraid of what the reaction might be. Well, as you can imagine, the variety of personalities in your clients and the separation experiences they have had will result in an equal variety of reactions to the news that you are leaving.

Some may feel abandoned. After all, you are leaving because of your school calendar, not because they don't need you, so they may wonder how much you really cared in the first place (Baird, 2002; Stanziani, 1993). Some may feel angry and believe their trust has been betrayed, even if you have been clear from the beginning that you will be leaving at the end of the semester. It is important to remember that whatever reaction you receive may be only part of the picture (Penn, 1990).

Termination, like the end of any significant relationship, usually creates feelings of anger, sadness, and appreciation. Your client may be expressing only one of those feelings, and indeed that may be the only feeling he or she can access, but it is important for you to remain open to the other feelings as well, and perhaps to help your client be open to them, too. It is also important to separate what clients choose to show you from what their deeper feelings may be.

Sometimes these feelings are expressed indirectly, not unlike your own feelings about ending the internship (Penn, 1990). In some helping fields, this is called *resistance* (Kiser, 2000). Some clients may be genuinely indifferent to your leaving. But others will feign indifference or be indifferent because their real feelings are too hard to accept. Another indirect way of dealing with termination issues is to demand that termination come immediately. In these cases, once you tell clients you will be leaving, they may stop coming to the agency or in residential settings ask for another worker to be assigned right away. Other clients may begin to exhibit older, more problematic behaviors, as if to say, "You can't leave me; I still need you." Finally, clients may begin to devalue the work you have done together, since it is easier to say good-bye to something or someone who, after all, wasn't that important anyway.

For most interns, saying good-bye to clients is an emotional experience for them as well, sometimes in unexpected ways. You may find yourself feeling guilty that you haven't been able to do more for some or all of your clients, and again, this feeling is often unrelated to how much you have actually accomplished (Baird, 2002). In other cases, you may worry that you are the only one who can work successfully with a particular client or group of clients. It may be that you are the first person to really get through to a client. That is a wonderful feeling, but it places extra weight on the termination process. As one intern said, "They do not receive the same help from others in certain areas.... It just makes me sick inside." Remember that although the client may regress some when you leave, chances are that at least some of the progress you have made will remain. It can, however, be hard to remember or believe that in the moment. If a client starts to regress, you may begin to question the value of your work or your effectiveness in general (Gould, 1978).

All human service workers are vulnerable to these reactions, but as an intern, especially if this is your first intensive experience in the helping field, you may be especially vulnerable. The closeness that can happen between worker and client is a heady, heartwarming experience, and it is especially affecting the first time (Baird, 2002). Both of us recall with special vividness our first few clients and the special difficulties that we faced in saying good-bye.

There are some pitfalls to watch out for as you take in and process clients' reactions to your leaving. It is easy to personalize these reactions, as if they were statements about *you*. Of course, they have something to do with you, but remember that you may also be hearing expressions of their current struggles with themselves and echoes of old wounds. It is important to separate your clients' needs and issues from your own. If, for example, you struggle with a need to be liked, then you may be especially vulnerable to a client who turns on you in anger or devalues the experience. It is important to deal with these feelings and issues separately from the decisions you make about how to respond to your clients. You may need to be liked, but liking you in that moment may not be the best thing for your client. If a client has never been able to express disappointment with anyone, then the ability to do that with you is far more important than your need to be reassured. Baird (2002) also cautions that some interns unconsciously try to get clients to make the termination easier for them. They seek reassurance that they have done a good job or that whatever problems remain are not their fault. In these times, it is especially challenging to keep the client's needs in the foreground. You may very well need some support and some focused attention to help you process your feelings and needs. That is what your instructor, supervisor, and peers are there for; don't turn to your clients.

Another pitfall is promising to come back and visit. Clients may ask you to come back or to maintain contact in some other way. They may be avoiding the difficulties of separation, they may not understand the boundaries of a professional relationship, or they may just be doing what they would do with anyone they have come to care about. Even if they don't ask, you may find yourself wanting to reassure them that you will visit or keep in touch. Use extreme caution before making any such promises. Consult with your supervisor and your faculty instructor. Above all, if you do make commitments, make them with care; the worst thing you can do is promise to keep in touch and then fail to do so.

The Future

The last issue to think about in termination with your clients is the future, both for them and for your relationship with them. Just as this is a time to review goals and accomplishments, it is also a time to set or reestablish goals for the future. You also need to think about who will be dealing with your clients after you leave. In many agencies, intern supervisors plan for this transition, and you should be part of this planning. If your supervisor is not raising this issue, we suggest that you do. This transition, often referred to as *transferring clients*, requires time and attention if it is to be done well (Baird, 2002; Rogers et al., 2000; Faiver, Eisengart, & Colonna, 1994).

The nature of the transfer will depend a great deal on your specific setting. You may actually have some clients that have been assigned to you. For example, there may be individuals who come to a clinic for counseling, a group you have been assigned to lead or co-lead, children and families whom you visit at their homes, or community groups that you meet with on a regular basis. In residential settings, it is not unusual for each resident to have a primary staff member, and sometimes interns move into this capacity after they have been there for a few weeks. In other settings, there will not be clients who have been officially assigned to you, but you may have developed especially close relationships with certain individuals, and it may be clear to you that you are the one to whom they turn most often. In any case, the issue of who is going to take over is one that should be discussed with your supervisor, with your coworkers, and of course with your clients. You should also be sure to reserve some time to work with the person who will be taking over, and you should begin this process well before your last day.

What sort of relationship will you have with your clients after you leave? What sort should you have? In many cases, the answer to both questions is "None." At most, you might keep in touch by mail. But for others, either because the client asks for more or because you want more, the issue becomes less clear. As we discussed in an earlier chapter, sexual relationships with clients while you are an intern are unacceptable and are a clear violation of the ethical standards of virtually every professional organization. Social relationships with clients are also discouraged, although in some human service settings they are unavoidable and even beneficial. Now, however, you are leaving, and there may be clients with whom you want to pursue a friendship or a romance. Some of you, particularly those working with adolescents, may find this absurd and out of the question. But others of you are working with clients your own age, and trust us, the issue does come up.

This issue is complicated once again by the wide variety of placements and clients that interns deal with. Post-termination relationships have been the subject of a good deal of discussion in the counseling field. For example, Salisbury and Kiner (1996) conducted a survey of attitudes about friendship and sexual relationships with former clients. Of the counselors surveyed, 70 percent said they thought a friendship was acceptable, while 30 percent thought a sexual relationship could be acceptable. However, they also advised a lengthy waiting period, averaging 25 months for friendship and 62 months for romance. So if you were thinking that maybe you could just pick right up with a different kind of relationship with one of your clients, you may want to think again. Herlihey and Corey (1992) have provided a thoughtful review of these and many other issues about relationships with clients.

The major concern about post-termination relationships is the unfair power differential between you and your former client (Salisbury & Kiner, 1996). Through working with clients, human service workers often have knowledge of their most sensitive issues and areas of vulnerability. Clients, on the other hand, usually have no such knowledge of their workers, and this puts them in a vulnerable position in a friendship or romance.

However, not all of you are counselors, nor are you necessarily working with clients who have come to you or the agency because of psychological or emotional issues. If you are working in a recreation center, doing community organizing, or providing health education for college students, the power differential referred to may be quite different. So unfortunately, there are no hard and fast rules to follow, just some important issues to think about. If you are considering pursuing any sort of relationship with one of your clients, you ought to talk that over very carefully with your supervisor and your campus instructor. Whatever you decide, remember that even though these people may no longer be your clients, the primary concern in your decision making still must be their welfare, not yours.

COMING TO CLOSURE WITH YOUR SUPERVISOR

For most interns the supervisor is the person who has most powerfully affected the experience. Your supervisor is in a position of power, at least with regard to you, and often a position of leadership in the agency. Usually the supervisor is the person with whom you have worked most closely, shared the most with, and learned the most from. Now it is time to close out that relationship, or at least move it to a new phase. There are two parts to this process: the formal and the informal (Berg-Weger & Birkenmaier, 2000). In the formal portion of the process you will have your final evaluation from your supervisor and perhaps you will evaluate your supervisor as well. In the informal—and equally important—phase, you will process and discuss the relationship you have had with your supervisor and your feelings about the ending of that relationship.

The Final Evaluation

You have probably had evaluations at your internship before this point; at least we hope so. Perhaps you are less nervous about them than you once were; perhaps not. In any case, you have one more to go through, and this can be the most important one of all. Many internship programs and placement sites use a written final evaluation of some sort. Most of them have their own format, so we are not going to provide one here. However, if you would like to see some examples, we refer you to work by Wilson (1981), Stanton and Ali (1994), Baird (2002), Horejsi and Garthwait (2002), and Berg-Weger and Birkenmaier (2000). As was the case with the initial or mid-semester evaluation, we recommend that you become familiar with the form used at your placement or reach an agreement about what it will be before the evaluation is actually completed. That way you will have a better idea what to expect.

Regardless of what form you use, or whether you use one, it is important that you have a final evaluation conference with your supervisor. This conference can either precede or follow the written evaluation. Some of the time can be used to prepare for that evaluation or to go over it. In any case, be sure to schedule an adequate block of time, at least an hour (Baird, 2002; Faiver et al., 1994; Shulman, 1983). There may be lots of tempting reasons for both you and your supervisor to avoid this session. You may be nervous about the feedback; your supervisor may not be good at endings either; both of you may be pretty busy trying to wrap up projects or clients. This is one of those times when a formal, scheduled time—a ritual—will ensure that the task actually gets done.

This conference will undoubtedly be more productive if you spend some time preparing for it, and we suggest several steps in doing so. First, before you even consider what is on the form, take some time to review the internship experience with all its joys and frustrations. You have been reflecting on a regular basis, but it is easy even so to miss some important moments or broader themes, and this is a time to do some overall reflecting (Baird, 2002; Berg-Weger & Birkenmaier, 2000; Kiser, 2000). Reread your journal and make some notes for yourself. If you had a mid-semester evaluation, look it over as well.

Stanton and Ali (1994) have suggested dividing your reflection into two sections: work performance and learning. As you will see, these two areas have some overlap, but they are worth considering separately. In discussing work performance, you and your supervisor should cover the areas where you seem to have been especially effective, the clients with whom you have worked most successfully, and the service areas where you have demonstrated the most skill (Faiver et al., 1994). Both of your perceptions of these issues are important. In discussing what you have learned, you should use your written goals and objectives as a reference. Discuss whether each one has been met, not met, or partially met, the reasons why, and the ways that what you have learned will be useful to you in your professional and personal life. If you are stuck, here are some topics from Berg-Weger and Birkenmaier (2000) that you may find helpful:

- Knowledge and skills you gained
- Areas of personal and professional growth
- New or confirmed areas of interest
- Goals met, unmet, and partially met
- Your level of confidence
- Your ability to function as part of a team
- Your ability to make good use of supervision

Another way to reflect on your experience is suggested by Brian Baird (2002), who says that there are both positive and negative lessons to be learned about clients, systems, coworkers, human problems, and yourself. For example, you may have learned that some people are in the helping professions for all the wrong reasons. On the other hand, you may have learned that there are quiet, heroic efforts being made day in and day out by human service professionals. You may want to try this exercise for yourself.

It is also important to prepare for the affective dimension of this experience (Stanton & Ali, 1994). Being evaluated is always an emotional experience to some degree, and for some interns the emotional stakes are very high. Brian Baird (2002) has suggested completing a self-evaluation, perhaps using the agency form, in which you downplay your strengths and call more attention to your areas for improvement. As you do so, monitor how you feel and how satisfied you are with your reaction. We also suggest you spend some time thinking about your emotional reactions to praise and criticism, as you did earlier. Some interns have a difficult time listening to criticism; others chafe at praise. Still others find it hard to tell other people how they really feel about them. Anything you can do to help yourself participate honestly and openly in this process will pay dividends.

During the evaluation itself, it is important to balance the positive and the critical (Baird, 2002; Shulman, 1983; Wilson, 1981). It is normal to want to focus on the positive and bask in the glow of your accomplishments and your supervisor's praise, but constructive criticism is just as important. There are bound to be things you need to work on, and perhaps areas where you didn't perform as well as you or your supervisor would like. Just as your clients need to know where they still have work to do, so do you. Ignoring these areas is a little like saying, "I'm perfect. I learned everything I need to know to be a professional in this field in just a few short weeks." That sounds more like a quote from a late night TV advertisement than the stance of a reflective, thoughtful professional. In come cases you may need to be assertive about asking your supervisor to identify areas for growth.

It is also important, during the session, to listen as carefully and critically as you can, clarify things if you need to, and ask questions (Berg-Weger & Birkenmaier, 2000). Feel free to ask for explanations or examples. If you don't understand what your supervisor is saying or on what it is based, it will be hard to learn from the positive or the negative.

Some of what you hear you may not like. Some you may not think is fair. It is important, though, to try and keep those two issues separate (Rogers et al., 2000). We suggest you listen quietly and respectfully, ask for clarifications and examples, and then go away and think about it for a while before responding. Take some time to vent your feelings, and then consider whether some or all of the criticism is valid. Try to reframe the criticisms into opportunities to learn and grow and remember that this is a professional issue. It is not a personal attack nor a comment on your overall worth and value (Munson, 1993). If, after careful reflection, you believe that comments have been made about you that are not true or not grounded in facts, you should discuss this with your campus instructor.

The end of an internship may also be a time to offer some feedback to your supervisor (Baird, 2002; Rogers et al., 2000; Faiver et al., 1994). Some supervisors will request this from you, and they may or may not give you advance notice. There are even forms available to offer written feedback (Rogers et al., 2000). You probably have mixed feelings about this prospect, and we don't blame you. On the one hand, this is an opportunity to tell your supervisor what went well for you in the relationship and where there may be areas for improvement. Both you and your supervisor may learn more about yourselves from this conversation (Kiser, 2000). You also have an obligation to future interns (Rogers et al., 2000). On the other hand, it is important to think carefully before offering criticism, regardless of how constructively put, unless it is requested (Baird, 2002). After all, your supervisor has and may continue to have power over you in the form of grades, letters of recommendation, and word of mouth in the community. Your campus instructor may be able to help you decide what and whether to share with your site supervisor. If you decide to give some feedback, we encourage you to practice or role-play, and to remember the principles of effective feedback.

The informal part of the process with your supervisor is both easier and more diffi-cult. For better or worse, there is no form to fill out, no big meeting, and no high-stakes evaluation. Just the world of feelings. For most interns, the relationship with a supervi-sor is among the most significant at the internship. For many, the relationship has been close and positive, both intellectually and emotionally. For others, the relationship has not been as close or satisfying, but still one in which a lot was learned. Now that the relationship is ending, and just as in your relationship with clients, the ending can engender a variety of feelings for both of you. If your relationship with your supervisor has been mostly an intellectual, dispassionate one, it may make saying good-bye easier (Haas, 1990). Some supervisors will initiate a conversation with you about these feel-ings; some will avoid it; and some will engage but need to be prompted by you. If pos-sible, it is important to try and give voice to your feelings.

Some interns find these conversations disappointing (which does not mean they are not important). Many interns develop powerful feelings of closeness, respect, and affection for their supervisors. In some cases, the feelings are mutual. But remember that while this is a new and unique experience for you, it may not be for your supervi-sor (Baird, 2002). It is more typical that the supervisor has enjoyed working with you and watching you grow, and will be sorry to see you leave, but does not feel personally close to you and has no desire to maintain a friendship after the internship is over. If this is the case, try not to let it change the value of your experience, or even of your feel-ings about your supervisor.

One last issue to consider is a letter of recommendation. If you have had a good experience, you will probably want a letter from your supervisor, either for your next internship or for future employment. It's a good idea to get it now, while you and the internship are still fresh in the supervisor's mind, rather than call or write months later. Baird (2002) offers some important guidelines to follow when asking for a recommendation:

- Before requesting the letter, ask if your supervisor feels comfortable writing a sup-portive letter for you. This may seem silly, especially if you have a good relation-ship. Many supervisors, if they do not feel they can write a positive letter, will simply suggest that you get someone else. However, you don't want to take the chance that the letter will be less positive than you expected, especially if it is sent directly to a school or employer.

- Make clear what your future goals are. A letter for graduate school may look some-what different than a letter for employment. There may also be some jobs for which your supervisor feels you are better qualified than others.

- Give plenty of notice, and let your supervisor know if there is a deadline.

- Provide your supervisor with whatever forms and envelopes are needed. Pre-addressed and stamped envelopes show courtesy and consideration.

SAYING GOOD-BYE
TO THE PLACEMENT

The last week of an internship is hard in many ways, as captured in this quote from a journal: "I think to myself this is the last time I am going to be doing this and seeing these people. It's been a busy and scary week." And now it is your last day, a day you have been looking forward to with an incredible mixture of feelings for weeks. You have had your final meeting with your supervisor, and you have prepared your clients for your departure. You imagine what it will be like when everyone says good-bye to you and what you will say as you leave. As you move through the day, though, no one says a thing to you about leaving. Your supervisor doesn't seem to be around, and for your coworkers, it seems to be business as usual. Conversations focus on the work, and during quiet times, it's the usual office small talk.

You go about your business, feeling a little confused but certain that someone will say something soon. Maybe they're being sneaky, and there is a little party planned for you later. No such luck. As you are leaving for the day, you say good-bye to people, pretty much as you always do. One of your coworkers looks up from what she's doing and says, "See you next week." "Well, no," you respond. "Today is my last day." "Oh my gosh," says your colleague. "I totally forgot! Well, hey, it's been great. Good luck to you." You start to reply, but the phone rings, a client calls, or a crisis erupts, and your colleague swirls away, back into the normal pace of life at the agency.

As you are driving home, you feel differently than you thought you would. Yes, there is joy and some relief, but also some hurt and a vague sense of emptiness. You try not to focus on that feeling, but it gnaws at you. What's wrong with those people anyway? Maybe you didn't mean as much to them as you thought. Maybe they're just rude and not quite the people you thought they were. You feel as if you deserve better from them after all you have done.

In a way, you're right. The events of the day have left you without a sense of appreciation and have interfered with your efforts to come to closure. Working with people for weeks and months and then forgetting their last day and not saying anything to them is perhaps insensitive. On the other hand, think about the pace of life at your placement. Most human service agencies are understaffed and extremely busy. In residential settings, large and small crises erupt all the time. And it's not the last day for anyone else. It may not even be the end of their day or week. So it is understandable that they might forget about your departure.

The point here is that you may need to be proactive in assuring you get that sense of closure you want. This is another time when a ritual may be in order. Some agencies have fairly elaborate rituals when someone leaves. There is a group meeting with clients, time set aside at a staff meeting, a party, or a farewell lunch. Other agencies don't do any of these things. Talk with your supervisor about the best way to mark the end of your internship. Ask what the norms are, and be clear about what you need. It may be that there is just no time for a group activity, but at least you will know that, and you can schedule individual 15-minute sessions with some of your coworkers. In our experience, most supervisors are glad to help you make something happen, but they might not think of it on their own or may want to know what you prefer.

Even if there is going to be a formal good-bye celebration, there may be individual coworkers to whom you have grown especially close. They may take the initiative to have a final conversation with you, or they may not for any number of reasons. Here again, you need to take the initiative and schedule some time with them. This is yet another opportunity to practice your feedback skills; it is a time to let them know specifically what you have learned from them, what you appreciate about working with them, and how you feel about them. They may, in turn, do the same for you. Remember, though, the main idea is for you to say what you want and need to say, and if you do, the conversation is a success. You don't want to go away with that nagging feeling that you wish you had said such and such to so and so. Whatever you get back from them is an extra bonus.

PUTTING TOGETHER A PORTFOLIO

Many interns we have worked with have found value in creating a portfolio of their internship experience. It is a reflective tool, but also a powerful asset as you seek employment or further education. There are a number of pieces you can consider including: reflections on your work, sample projects, case studies (being mindful of confidentiality), performance reviews, and so on. Whatever time you spend on this endeavor will be time well spent, but if you are going to invest serious energy in this work we suggest you follow a process for portfolio development (King & Sweitzer, 2002):

- **Consulting with Peers** Work in small groups without an instructor. Exchange ideas about your portfolios and begin to sketch them out.
- **Constructing the Vision** Here is where you begin to think about the contents, although final decisions do not come until later. Write a table of contents and an introductory statement. Make a list of potential contents. Think about how you would like to package the portfolio.
- **Creating the Vision** Two steps here. In the first, you brainstorm all the possible organizing lenses or metaphors you might use to create and organize your portfolio. Examples include humanitarian student, emerging professional, ethical decision maker, and caring advocate. Then begin collecting the documents you need.
- **Communicating the Vision** Here is where you choose the actual format and design of the portfolio.

LOOKING FORWARD

Now your internship is over, and your future has begun. Probably, there were times when you thought this day would never come, or times when you were amazed and unnerved at how fast it was approaching. New challenges await you, of course. Some of you may be headed for another internship, some for a job in the field, and still others for the next level of schooling and internships. The helping professions, and the fields of social and public service, are rich and varied environments, with countless opportunities for you (Diaz, 2002).

It is important to remember that the stages of an internship are not experiences you go through just once. In your new job or field placement, you are going to go through them again. The concerns about expectations and acceptance and the challenges of keeping yourself moving and growing, of confronting problems, and perhaps of ending well will all visit you again and again. When we tell our students this piece of news, some of them roll their eyes and hold their heads. The good news is that although you will go through the stages again, it will not be the same. You have learned a great deal, and that learning will go with you.

You have learned valuable skills. If you continue to use them, they will grow sharper and more integrated over time. Even if your life and career change directions now, you can still take a lot of what you have learned with you. You have learned what to expect from an internship. The challenges will have a different shape and pace because they will be happening in a different place. You have grown and changed so much that it may feel like these challenges are happening to a different person. And of course, you will handle the issues in a new way. However, the concerns themselves should seem familiar to you if you can get enough distance and perspective from your everyday activities. Perhaps you will continue to keep a journal and pause for regular reflection on what you have written.

You have also learned a great deal about yourself. We have tried to emphasize that sort of learning in this book, and interns often tell us that this is the most powerful learning of all. You know more about how you respond to challenges and why. And you know that self-understanding is a process, not an accomplishment. You have more tools to pursue your self-understanding, and the practice you have had, if you continue it, will make those tools more and more second nature to you.

It may seem odd for us to say this, since we have never met you, but we wish you well. We are always glad to hear from former students and would be glad to hear from you about your continuing journey and how this book may or may not have been useful.

We leave you as we began, with two quotes from student journals:

I was given the opportunity to prove to myself that I could do it. This alone has allowed me to feel competent. I tested out my skills and got a professional feel about them. Now I have the key in my hand. I feel ready to move on. I am still not quite sure which doors this key will open, but I am sure that whatever I face I will deal with as best I know how.

It's like moving out of your house. You're leaving something that has had such an impact on your life and starting something unknown and exciting. You are leaving a foundation of growth and a place filled with memories. You're going into unknown territory, and that is exciting.

For Further Reflection

FOR PERSONAL REFLECTION:
SELECT THOSE QUESTIONS MOST MEANINGFUL TO YOU

1. What is on your agenda right now? What projects need to be finished at your internship? How about other classes? What else do you have to take care of in the next few weeks?

2. Think back to times in your life when you faced endings. What was it like for you then? What feelings did you have? How did you handle the ending?

3. Can you make any general statements about how you tend to handle good-byes? Are you satisfied with your patterns and tendencies in this area?

4. What will be most difficult in ending your internship? What can you do to ensure that you end the way you want to?

5. How will you approach termination with your clients? When will you tell clients you are leaving? How can you come to some closure with them? What future goals do you have for them? How do you feel about saying good-bye?

6. Think for a moment about your supervisor. What are some of the things you have learned? What is it that you most appreciate? What do you wish had been different? Is there anything you want to say to your supervisor before you leave?

7. Are there other individuals at your placement that you want to be sure to say good-bye to? How will you do this?

8. What plans do you have to acknowledge or celebrate the end of your internship?

SPRINGBOARDS FOR DISCUSSION

1. Those of you who are working with clients may want to take some class time to focus on just how you will let them know you are leaving. This is another opportunity to role-play.

2. Take some time in class to focus on the issue of offering feedback to supervisors. Role-play these situations in triads, with an intern, a "supervisor," and an observer. The observer should use the principles of effective feedback to note strengths and weaknesses.

For Further Exploration

Baird, B. N. (2002). *The internship, practicum, and field placement handbook: A guide for the helping professions* (3rd ed.). Upper Saddle River, NJ: Prentice Hall.

This book contains a comprehensive, thoughtful discussion of a variety of issues associated with the ending of an internship.

Brill, N. I., & Levine, J. (2001). *Working with people: The helping process* (7th ed.). Boston: Allyn & Bacon.

Especially helpful in thinking about termination with clients.

Diaz, A. (2002). *The Harvard college guide to careers in public service*. Cambridge, MA: Career Services Publications.

Up-to-date discussion about a wide range of careers and opportunities.

Gould, R. P. (1978). Students' experience with the termination phase of individual treatment. *Smith College Studies in Social Work, 48*(3), 235–269.

An excellent discussion of the unique nature and demands of "forced terminations."

References

Baird, B. N. (2002). *The internship, practicum, and field placement handbook: A guide for the helping professions* (3rd ed.). Upper Saddle River, NJ: Prentice Hall.

Berg-Weger, M., & Birkenmaier, J. (2000). *The practical companion for social work: Integrating class and field work*. Needham Heights, MA: Allyn & Bacon.

Brill, N. I., & Levine, J. (2001). *Working with people: The helping process* (7th ed.). Boston: Allyn & Bacon.

Diaz, A. (2002). *The Harvard college guide to careers in public service*. Cambridge, MA: Career Services Publications.

Fair, S. M., & Bressler, J. M. (1992). Therapist-initiated termination of psychotherapy. *Clinical Supervisor, 10*(1), 171–189.

Faiver, C., Eisengart, S., & Colonna, R. (1994). *The counselor intern's handbook*. Pacific Grove, CA: Brooks/Cole.

Gould, R. P. (1978). Students' experience with the termination phase of individual treatment. *Smith College Studies in Social Work, 48*(3), 235–269.

Haas, S. (1990). *Hearing voices: Reflections of a psychology intern*. New York: Penguin.

Herlihey, B., & Corey, G. (1992). *Dual relationships in counseling*. Alexandria, VA: American Counseling Association.

Horejsi, C. R., & Garthwait, C. L. (2002). *The social work practicum: A guide and workbook for students* (2nd ed.). Needham Heights, MA: Allyn & Bacon.

Kegan, R. (1982). *The evolving self: Problem and process in human development*. Cambridge, MA: Harvard University Press.

King, M. A., & Sweitzer, H. F. (2002). *Reflection as keystone: Growth and learning in service*. Annual conference of the National Organization for Human Service Education, Providence, RI, October.

Kiser, P. M. (2000). *Getting the most from your human service internship: Learning from experience*. Belmont, CA: Wadsworth.

Munson, C. E. (1993). *Clinical social work supervision* (2nd ed.). New York: Haworth Press.

Penn, L. S. (1990). When the therapist must leave: Forced termination of psychodynamic psychotherapy. *Professional Psychology: Research and Practice, 21*, 379–384.

Rogers, G., Collins, D., Barlow, C. A., & Grinnell, R. M. (2000). *Guide to the social work practicum*. Itasca, IL: F. E. Peacock.

Salisbury, W. A., & Kiner, R. T. (1996). Post termination friendship between counselors and clients. *Journal of Counseling and Development, 74*(5), 495–500.

Schafer, R. (1973). The termination of brief psychoanalytic psychotherapy. *International Journal of Psychoanalytic Psychotherapy, 11,* 135–148.

Shulman, L. (1983). *Teaching the helping skills: A field instructor's guide.* Itasca, IL: F. E. Peacock.

Stanton, T., & Ali, K. (1994). *The experienced hand: A student manual for making the most of an internship* (2nd ed.). New York: Caroll Press.

Stanziani, P. (1993). *Practicum handbook: A guide to finding, obtaining, and getting the most out of an internship in the mental health field.* Cambridge, MA: Inky Publications.

Suelzle, M., & Borzak, L. (1981). Stages of fieldwork. In L. Borzak (Ed.), *Field study: A sourcebook for experiential learning* (pp. 136–150). Beverly Hills, CA: Sage Publications.

Watzlawick, P., Weaklund, J. H., & Fisch, F. (1974). *Change: Principles of problem formation and problem resolution.* New York: Norton.

Wilson, S. J. (1981). *Field instruction: Techniques for supervisors.* New York: Free Press.

Index

Note: page numbers followed by *n* refer to footnotes